DATE DUE

No card

948.5
ZIC
    Zickgraf, Ralph.
    Sweden

Adirondack Central School
Boonville Middle Library

‹ SWEDEN ›

MAJOR WORLD NATIONS
SWEDEN

Ralph Zickgraf

CHELSEA HOUSE PUBLISHERS
Philadelphia

Chelsea House Publishers

Contributing Author: Tom Purdom

Copyright © 1999 by Chelsea House Publishers,
a division of Main Line Book Co.
All rights reserved.
Printed and bound in the United States of America.

First Printing

1 3 5 7 9 8 6 4 2

Library of Congress Cataloging-in-Publication Data applied for

ISBN 0–7910–4749–0

CONTENTS

Map 6
Facts at a Glance 9
History at a Glance 11
Chapter 1 Sweden and the World 17
Chapter 2 The Northern Lands 21
Chapter 3 Stockholm and the South 39
Chapter 4 Early History 51
                Color Section Scenes of Sweden 57
Chapter 5 Building the Swedish State 69
Chapter 6 Government and Economy 85
Chapter 7 The People and Culture 99
Chapter 8 Sweden and the Future 111
Glossary 115
Index 117

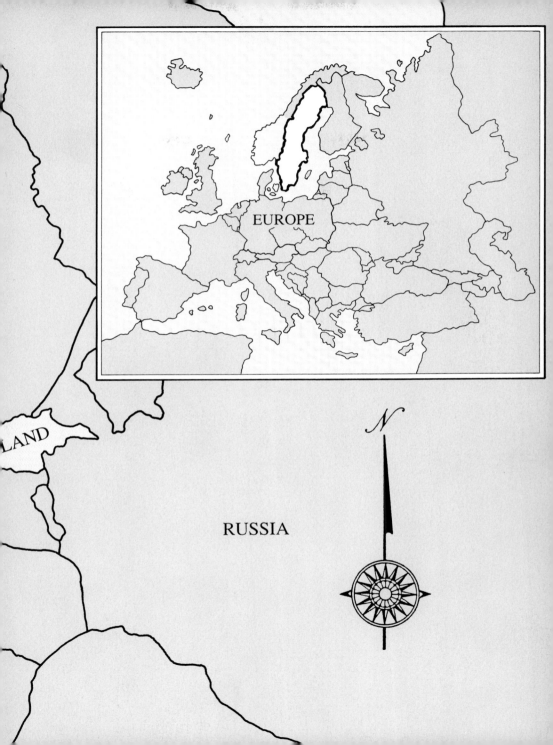

◄ FACTS AT A GLANCE ►

Area	173,732 square miles (449,964 square kilometers)
Highest Point	Mt. Kebnekaise, 6,926 feet (2,111 meters)
Climate	Temperate in the south to arctic in the north; temperatures in the south from 30° F (−1° C) in February to 63° F (17° C) in July; temperatures in the north from 7° F (−14° C) in February to 64° F (18° C) in July; annual precipitation from 23 inches (585 mm) in the south to 15 inches (380 mm) in the north
Population	8,842,000
Population Density	51.0 people per square mile (19.7 per square kilometer)
Population Distribution	Urban, 84 percent; rural, 16 percent
Capital	Stockholm (population 718,462)
Major Cities	Gothenburg (population 454,016), Malmö (population 248,007), Uppsala (population 184,507)
Official Religion	Church of Sweden (Lutheran), 94 percent
Other Religions	Roman Catholic, 1.5 percent; Pentecostal, 1.0 percent; other, 3.5 percent
Ethnic Groups	Swedish, over 80 percent; foreign-born or first-generation immigrants (such as Finns, Danes, Norwegians, Greeks, and Turks), approximately 12 percent

Economy

Major Products	Iron and steel, precision equipment, wood and paper products, processed foods, automobiles
Major Resources	Zinc, iron ore, lead, copper, silver, timber, uranium, hydropower potential
Makeup of Gross Domestic Product	Agriculture, 2 percent; industry, 27 percent; services, 71 percent
Currency	Krona (plural kroner), divided into 100 ore

Government

Form of government	Constitutional monarchy
Formal head of state	King
Head of government	Prime minister
Voting rights	All citizens age 18 and older
Legislature	Unicameral parliament, called the Riksdag

HISTORY AT A GLANCE

3000 to 2000 B.C.	Stone Age hunter-gatherers settle in Sweden.
2000 to 1500 B.C.	Bronze Age Boat-Axe people invade Sweden and begin trade with nations near the Mediterranean Sea.
100 B.C.	The Iron Age begins in Sweden.
800 to 1100 A.D.	Swedish Vikings explore and settle in the Dnieper and Volga valleys.
829	The first Christian missionary reaches Sweden.
1006	Olof Skötkonung becomes the first Swedish king to be baptized as a Christian.
1130 to 1250	Rival groups of nobles engage in civil war.
1156 to 1160	King Eric, later Saint Eric, reigns. He annexes Finland to Sweden.
1250	Birger Jarl begins the Folkung dynasty, which rules until 1374.
1252	Stockholm is founded as a fort on Lake Mälar.
1347	A royal charter to mine copper at Falun is granted to Stora Kopparberg, the world's oldest industrial firm.
1394	The Kalmar Union unites Sweden, Norway, and Denmark under the rule of Denmark's Queen Margaret.
1434	A peasants' rebellion led by Engelbreckt Engelbrecktsson leads to the establishment of the Riksdag, the first Swedish parliament.

1477	Sweden's first university is founded at Uppsala.
1520	The Stockholm Bloodbath occurs when King Christian II of Denmark executes 80 Swedish noblemen in Stockholm.
1527	The Lutheran Church becomes Sweden's official church.
1521 to 1523	Gustav Vasa drives Danish armies out of Sweden, dissolves the Kalmar Union, and is elected king of Sweden.
1619	King Gustavus II Adolphus founds the port city of Gothenberg
1638	Sweden founds the colony of Nya Sverige in North America.
1655	Swedish traders establish a trading post on the Gold Coast of Africa.
1658	Sweden wins Halland, Skåne, and Blekinge from Denmark.
1700	King Charles XII defeats the Russian army of Peter the Great at Narva.
1707	King Charles XII invades Russia.
1708	Peter the Great defeats Charles XII at the Battle of Poltava.
1718	Charles XII dies in battle in Norway. Sweden loses all its overseas possessions except Finland.
1719	A new constitution gives more power to the Riksdag and limits the king's power.
1772	King Gustav III seizes control of the government and rules as an absolute monarch.
1792	Gustav III is assassinated. He is succeeded by his son, Gustav IV.
1805 to 1809	Sweden, drawn into the Napoleonic wars, is driven out of Finland.

1809	Gustav IV is deposed. His uncle is crowned King Charles XIII, but a new constitution limiting the power of the monarchy is accepted.
1810	Jean Bernadotte, marshal of France, is made crown prince of Sweden. He takes the name Charles John.
1814	After a brief war with Denmark, Sweden signs the Peace of Kiel. The Danish gain control of Norway.
1818	Bernadotte is crowned King Charles XIV.
1866	Alfred Nobel invents dynamite.
1860 to 1890	Successive bad harvests and growing unemployment result in massive emigration to the United States.
1870 to 1900	Swedish industry begins explosive growth. People begin to move to the cities.
1896	The Nobel Prizes and the Nobel Institute are established.
1905	Sweden grants Norway independence.
1914	Sweden declares its neutrality in World War I.
1919	Labor unrest and fear of communist revolution lead to political and social reforms. Universal suffrage is established.
1932	Sweden feels the effects of the worldwide depression. Widespread unemployment and economic scandal lead to the election of the Social Democratic party.
1939 to 1945	Sweden declares its neutrality during World War II. Although it is the only Scandinavian country not occupied by Germany, it is forced to allow German troop movements through the country.
1946	Sweden joins the United Nations.

A Swedish girl of Dalarna, in her native costume, picks flowers in a meadow.

1946 to 1976	Under Social Democratic rule, the country continues economic growth and political and social reform.
1976 to 1982	The world-wide oil crisis causes a slowdown in economic growth and a rise in unemployment. For six years, a succession of nonsocialist governments struggle with the economy and the issue of nuclear power.
1980	In a national referendum, the Swedish people decide to stop building nuclear reactors and to phase them out by the beginning of the 21st century.
1982	The Social Democrats resume power, under Prime Minister Olof Palme.
1986	Olof Palme is assassinated. Social Democrat Ingvar Carlsson succeeds him as prime minister.
1991	Conservative coalition, pledging to lower taxes and make cuts in welfare state, wins election.
1994	Social Democrats return to power as leaders of coalition government.
1995	Sweden joins European Union (EU).
1997	Government begins to phase out nuclear power plants.

These apartment houses in Stockholm are built in the typical modern Swedish architectural style.

Sweden and the World

Sweden is a land of startling contrasts. Twelve hundred years ago, tall blond Swedish Vikings in dragon ships spread terror and death throughout Europe. Today their Swedish descendants work as United Nations peacekeeping troops to promote world peace.

Sweden has been one of the world's leading producers of weapons—yet the country has been at peace since 1815. Alfred Nobel, a Swede, invented and made a fortune from the sale of dynamite. But he left that fortune to establish the Nobel Prize, which honors and rewards excellence in science, literature, and the pursuit of peace.

In the 19th century, Sweden was so poor and overpopulated that almost 20 percent of the country's population left to avoid starvation. Today Sweden's per capita (per person) wealth is the highest in Europe and one of the highest in the world.

Swedes love nature and solitude, yet 84 percent of the population lives in cities with closely packed high-rise apartment buildings. On the same day, in different parts of this California-sized country, a visitor can find a fur-clad Lapp milking reindeer and a white-smocked technician inserting fuel rods into a nuclear reactor.

What can we learn about this Land of the Midnight Sun, this welfare state with the thriving capitalist economy?

Sweden's climate, culture, and history have been shaped by geography. Set in the far northern corner of Europe, Sweden was long protected from domination by other European powers by its isolated location and the imposing Baltic Sea. At the same time, Sweden was close enough to the rest of Europe to take part in war, trade, or politics with the European nations.

Disastrous involvement in European power struggles in the 18th century helped to slow Sweden's economic development, so that it became one of the poorest countries in the world. Once the Industrial Revolution began in the 19th and 20th centuries, Sweden's abundant natural resources helped it leap forward economically. Those resources include minerals, wood products, and many large, fast-flowing rivers to power hydroelectric plants.

The country's most valuable resource is its people. They are energetic, inventive, and noted for industry, cleanliness, and a love of order. They are sometimes criticized, by themselves as well as by

In Stockholm, a group of Swedish citizens gathers at the spot where Prime Minister Olof Palme was assassinated.

their neighbors, for being rigid, condescending, and gloomy. That is hard to believe, though, when one sees them gaily, even wildly, celebrating the brief and glorious return of summer every year at Midsummer's Eve.

Today, Sweden's prosperity, peacefulness, and lovely landscape are threatened by dangers that cannot be kept at bay by distance and physical isolation. In 1986, the partial meltdown of the Chernobyl nuclear power reactor in the Soviet Union demonstrated that Sweden is no longer protected by its far northern location. Winds blew radioactive fallout from the reactor north, where rains washed it out of the sky and into the soil and streams of Sweden, Norway, and Finland.

In that same year, Sweden was hit with another shocking intrusion of modern reality when its prime minister was assassinated. Olof Palme was shot down on a Stockholm street as he walked with his wife from a movie. There were no bodyguards. Until that evening, very few Swedes would have dreamed that bodyguards were necessary.

Sweden's policy of armed neutrality worked during two world wars, and Sweden is determined to keep it working. Sweden made it through the entire Cold War without joining the North Atlantic Treaty Organization (NATO), even though its relations with the Soviet Union were strained. Yet Sweden has always understood that it is part of the international community. Sweden became a full member of the European Union in January 1995, and it participates in most of the other organizations that foster international cooperation. Sweden contributed to the forces that intervened in the former Yugoslavia in 1996. Swedish soldiers have also been an important component in the United Nations peacekeeping forces that have stood guard in trouble spots all over the world.

The midnight sun is a phenomenon of northern Scandinavian lands.

The Northern Lands

Sweden is located on the eastern side of the Scandinavian peninsula, which juts out from the northeast corner of the continent of Europe, above the Arctic Circle, and extends south for more than 1,000 miles (1,600 kilometers). The North Sea forms the western boundary and the Baltic Sea and the Gulf of Bothnia form the eastern boundary of the peninsula. To the south, two narrow arms of the North Sea, the Skagerrak and the Kattegat, separate the Scandinavian peninsula from Denmark. Denmark is located on another peninsula that extends north from the European mainland. On a map, the two peninsulas look as if they might fit together like jigsaw puzzle pieces.

Sweden shares the Scandinavian peninsula with Norway; the border between their two countries is 1,044 miles (1,619 kilometers) long. Across the Gulf of Bothnia lies Finland, which shares Sweden's 355-mile (590-km) northeast border. Six hundred miles west of Norway, isolated in the cold North Atlantic, is the island nation of Iceland. Sweden, Norway, Finland, Denmark, and Iceland make up the region called Scandinavia.

With a total land area of 173,732 square miles (449,954 square kilometers), Sweden is the largest Scandinavian country and the fifth-largest country in Europe. Sweden can be divided into four regions. The northern two-thirds of the country are an area of rugged mountains and forests called Norrland. South of Norrland is Svealand, a less mountainous region named after the Germanic tribe that originally settled the area. South of Svealand is the central highlands region of Gothland, named after the Goths, another Germanic tribe.

The fourth and most southerly region is Skåne, a flat fertile area where most of Sweden's agriculture is located. Skåne includes the two islands of Gotland and Öland. Skåne's terrain is gentler and its climate is warmer than most of Scandinavia. The Swedes say God gave them Skåne so they would know what Europe is like.

To most Swedes, the glory of the land lies in its lakes and forests, most of which are found in Norrland. More than 100,000 lakes and many rivers cover nearly 30 percent of the entire country. Seemingly endless forests cover about 50 percent of the land area.

The forest are mostly pine, and the wood, paper pulp, and other products from the trees make them one of Sweden's most important resources. The rivers have great economic importance, too. They are the highways over which the harvest of trees is brought to the saw mill, and their rushing waters are converted by thousands of dams and turbines into electricity.

The mountains of Norrland are highest near the northwest border with Norway. Mount Kebnekaise, Sweden's highest peak at 6,926 feet (2,111 meters), is in the county of Norrbottens. Norrbottens is Sweden's largest and least populated county. Located north of the Arctic Circle, it is a harsh land. Winter temperatures drop to $-13°$ Fahrenheit ($-45°$ Centigrade), and the annual precipitation is less than 15 inches. During the winter, the sun does not rise above the horizon for weeks at a time, and the days are only a few short hours of murky twilight.

In the summers, however, the sun is high in the sky and the days are long. All through Norrland it remains bright enough to read at 2 A.M., and people must put heavy curtains on the bedroom windows in order to sleep. In Karesuando, one of the most northerly permanent settlements in the world, the sun does not set at all between May 26 and July 18. Instead of going down in the west, the sun dips to the south, appears to roll along the southern horizon like a marble along the rim of a plate, and then rises into the sky again. This phenomenon is why Sweden and Norway are sometimes called Lands of the Midnight Sun.

The northern half of Norrland is also called Lappland, after the Lapp people, who have lived there since prehistoric times. Most Lapps were originally nomads who herded reindeer (nomads are people who do not live in permanent dwellings, but who move from place to place as the seasons change). The Lapps ate reindeer meat, drank reindeer milk, and used reindeer hides for clothing and tents. They and their herds spent the winters in the alpine forests of southern Lappland. In the summer, they moved their herds north to the tundra, the treeless plain of the Arctic that freezes solid in winter. During the summer the subsoil does not thaw, so the melted snow saturates the topsoil, making it black and mucky. With summer's long days of intense sunlight, the tundra produces a luxuriant carpet of tough dwarf herbs and shrubs that make excellent grazing for the reindeer.

Today only about 10 percent of Sweden's Lapps are nomads. Most live on small farms and supplement their crops with trapping or logging. In the Lappland village of Jokkmokk, the government has build a Lapp college. Educators hope to teach and train young Lapps to work in Sweden's modern economy of industry and services without losing their ethnic identity.

Norrland is Sweden's frontier. It has a smaller population and is less developed than the rest of the country. So it is surprising to see the modern office towers and apartment blocks of the city of Kiruna rising where the mountains and the tundra meet. Founded in 1898, Kiruna is Norrland's most populous city. Its 30,000 inhab-

A Lapplander shepherds his herd of reindeer.

itants make it the largest settlement north of the Arctic Circle. In one sense it is probably the largest city anywhere: Kiruna's leaders have declared that the city limits are the same as the borders of the township. As a result, Kiruna contains 5,000 square miles (13,000 square kilometers).

Kiruna is the site of the world's largest underground mine. Deep within the earth is the largest concentration of high-grade iron ore in Europe. The mines at Kiruna and at Gällivare, a smaller mining center about 60 miles (96 km) southeast, have produced over 16 million tons (14.5 million metric tons) of iron ore each year. Railroads take the ore west to the Norwegian North Sea port of Narvik, or southeast to the Swedish port of Luleå, which is located on the Gulf of Bothnia. Even though the gulf is closed by ice for at least three months every winter, Luleå remains a busy city because of the huge government-owned steel mill located there.

The border between Sweden and Norway, Norrland is the drainage divide for the Scandinavian peninsula: the mountains here force water on the Norwegian side of the border to flow west into the North Sea. Water on the Swedish side flows east into the Gulf of Bothnia. Norrland is striped diagonally with large rivers that flow out of the mountains and through the sweeping pine forests into the Gulf of Bothnia.

The Muonio and Torne rivers mark the northern border with Finland. Farther south the Skellefteälvan River roars through the province of Västerbottens. At the mouth of the river is Skellefteå, a busy seaport and mining center. This district is called the Gold Coast because of the gold and other precious metals that are mined here. The Indal River begins in the mountains of Jämtland, which form the border with Norway. The Jämtland Mountains are a popular vacation area. Åreskutan, Helagsfjallet, and Sylarna are some of the peaks that draw crowds of skiers in the winter. In the spring and summer, city dwellers spend their vacations here in cottages and cabins. Many Swedes take vacations in the mountains, along the coast, or on one of the thousands of *skerries* (small, rocky islets) along the coast.

The geography of Norrland shows how the Ice Ages shaped Sweden and the Scandinavian peninsula. Ice Ages were long periods when ice and snow in the mountains did not melt, but built up in huge glaciers. The glaciers grew in size and slowly moved south. The glaciers that buried Sweden were so heavy that the land actually sank. The coastline and the entire south of Sweden were under water for centuries. Thick layers of sediment settled on these areas while the land was under water. These sediments made the soil suitable for farming.

In the mountains, the glaciers scraped and scoured the land as they crept downhill and scooped up piles of stone, sand, and gravel. When the climate warmed, the glaciers stopped moving. The glacial ranges in the south melted, leaving behind the stone and gravel in piles and ridges that are called moraines and eskers. The eskers are a valuable source of construction material. The moraines formed dams, which backed up water in the hollows scoured out behind them to form the beautiful lakes that stretch along the base of Sweden's high peaks.

A rich variety of wildlife lives in the lakes and forests of Norrland. Wolf, bear, and lynx are now rare and are protected from hunting. Magnificent elk roam the forests, as well as deer, badgers, otters, and rabbits. There are songbirds, quail, and birds of prey, such as hawks. Along the coast are found many species of duck and geese, including the rare eider duck. Lordly cranes and herons nest in northern marshes. Until recently, Norrland's lakes and rivers teemed with trout, char, and salmon. Fish, however, have been vanishing from Norrland. Some Swedes believe the fish population is declining because of the effects of acid rain. Forests, too, have been affected by the acid rain.

Large herds of elk roam the forests of Sweden.

These logs on Lake Siljan will be used to make matches.

Acid rain occurs when rain and snow become polluted with sulfuric acid. The acid floats into the atmosphere from the tall smokestacks of electricity-generating plants and factories that burn coal or oil for fuel. When the polluted rain or snow lands in lakes, rivers, and forests, it kills the fish or trees because of the sulfuric acid it contains. Sweden was one of the first countries to feel the effects of acid rain, and it was one of the first countries to join with other nations to work toward stopping the spread of acid rain. Acid rain is truly an international problem, because the sulfuric acid can travel high in the atmosphere for hundreds of miles from its country of origin before falling to blight the environment.

The county of Gavleborgs is at the southern end of Norrland. This picture-book landscape of forests and lakes, rocky coastlines and gentle bays, is a favorite with tourists. Swedes have lived in this region for centuries, and the sense of tradition is shown by the many fine old houses and pleasant villages tucked into the hills. Every year, in July, the world's largest folk-dancing competition is held in Gavle, Hårga, Bollnäs, Arbrå, and Järvsö. Gavleborgs is where Sweden's rocky, remote frontier meets Sweden's heartland, the regions of Svealand and Gothland.

Svealand

Swedes have lived in Svealand and Gothland since the land was settled by their Iron Age ancestors, the Sveas and the Goths. (The Iron Age began about 800 B.C., when people began to make their tools and weapons from iron.) It was the Sveas who gave Sweden its name. In their language, *svea* meant simply "us," and *rike* meant "kingdom." Sverige—"our kingdom"—the modern Swedish name of the country, is a 2,000-year-old legacy from the Svea.

Svealand forms the central part of Sweden, a region of clay plains studded with forested hills. The white bark of birches and the flutter of aspen leaves make these woods more colorful than the vast dark-green pine forests of Norrland. There are many lakes, relics of the last Ice Age, when the land was under water. One of the loveliest of these lakes is 25-mile-(40-kilometer) long Lake Siljan, the "eye of Dalarna."

Dalarna ("land of dales, or small sheltered valleys"), in the northeast corner of Svealand, is one of Sweden's most historic regions. From Dalarna, 16th-century Swedes led the revolt for independence from Danish rule. Gustav Vasa, the leader of the revolt, was born here.

Many modern factories are tucked into the hollows, but farming is still an important way of life in Dalarna. People celebrate the lore and traditions of early farm life. In Rättvik, on the shores of Lake Siljan, an open-air museum houses centuries-old farmhouses and barns in which people demonstrate farm life. In nearby Leksand, 20,000 people meet each year to celebrate Midsummer's Eve with the raising of a 70-foot (21-meter) maypole. Many wear traditional peasant (farmer) dress for the celebration. The women wear finely pleated skirts, brightly colored aprons, and white caps. Men wear buckskin tights, green vests, leather aprons, and white jackets.

The largest town in Dalarna is Falun. Located on the southern tip of Lake Siljan, it is the site of Copper Mountain and of a mine that has been in operation for more than 700 years. During the 17th century, copper was in great demand for weapons, and the mine made Sweden the world leader in copper production. Stera Kopparbers, the company that was given a royal charter in 1347, still operates the mine, as well as iron mines, steel mines, and power stations throughout Sweden.

In 1687, a spectacular cave-in at one mountain mine near Falun left a crater big enough to hold a football stadium. Today, the hollow mountain is a tourist attraction. The mine museum nearby has examples of copper coins minted in the 17th century. One is 24 inches (60 centimeters) across and weighs 44 pounds (20 kilograms). In 1644, when it was minted, this huge coin would have bought a horse and wagon.

South of Lake Siljan is Lake Vänern. With an area of 2,156 square miles (5,584 square kilometers), it is the largest lake in Scandinavia and the third-largest lake in Europe. The surrounding land is called Värmland. Its geography varies from gentle, rolling fields in the east to steep hills and deep, narrow lakes near the Norwegian border. Swedes consider Värmland a special place because it was idealized in the novels of Selma Lagerlöf, who won the 1909 Nobel Prize for literature. Her birthplace, at Mårbacka, west of Lake Vänern, is a national shrine.

Near Lagerlöf's birthplace is Filipstad, where the grave of John Ericsson lies. Ericsson created a Värmland different from the wild and rustic country depicted in Lagerlöf's novels. A 19th-century inventor and industrialist, Ericsson is most famous as the man who built the *Monitor*, which won the first battle between iron-clad warships during the American Civil War. Värmland's abundance of easily mined iron ore and wealth of forest for fuel made the region an early center of steel production. Värmland is still a source of war machines: the international arms conglomerate Bofors is based here.

The small town of Uddeholm is headquarters of the giant Uddeholm Corporation, which began as a 17th-century ironworks, and today is the largest private (nongovernment) employer in Sweden. Nearby Munkfors is the site of one of the company's huge steelworks. The Klarälven River, which runs through the town, is peaceful, clean, and a prized vacation spot. The Swedes' love of nature and the care they have given their environment have resulted in the wonderful paradox of Värmland—an industrial center in a rural paradise.

Karlstad, at the northern end of Lake Vänern, is the capital of Värmland. Its population of about 80,000 people makes it the largest city in the area. Karlstad is one of the oldest cities in Sweden. It was formerly called Tingvalla, because it was the site of the *ting*, or Viking legislature. The city does not look old, however, because it has been rebuilt several times because of fires. The Göta River and the Trollhäte Canal connect Lake Avnern with the sea, making Karlstad one of the busiest freshwater ports in the world.

East of Värmland the plains and forested hills extend to the shores of Lake Mälar and beyond to the Baltic Sea coast. This is the most populated and industrially active part of Sweden. The city of Örebro and the nearby towns of Hällefors, Karlskoga, and Degerfors are centers of iron and steel production. The blackened remains of charcoal piles and ancient blast furnaces are found throughout this area.

At the eastern end of Lake Mälar is Stockholm, the capital of Sweden. Forty-five miles (73 kilometers) northwest is Uppsala, which once was the home of the kings of Svea. Throughout the surrounding province of Uppland are giant round mounds, called tumuli, in which the Svea kings were buried. On a hill overlooking the city is a magnificent castle, begun in the early 16th century by Gustav Vasa, the first king of united Sweden. Uppsala also has a very large cathedral, dating from the 15th century. Uppsala was a seaport during the Svea kingdom, and archaeologists have uncovered ancient ships buried in and around the city. The sea waters gradually receded, and since the 8th century Uppsala has been an inland city on the banks of the Fyrisån River.

Uppsala University, founded in 1477, is the oldest university in Sweden. Its most famous graduate is Carl Linnaeus, the 18th-century physician and naturalist whose two-name system of describing and naming plants became the basis for taxonomy—the science of classifying and naming plants and animals. Uppsala University houses the famous 4th-century silver Bible. This incomplete version of the New Testament in the ancient Gothic language was written on parchment in silver and gold ink. It is considered by some to be the world's most valuable book.

Ancient Uppsala coexists with modern Uppsala. Like most Swedish cities, Uppsala is ringed by large apartment complexes and modern highways. Ferry and boat service across Lake Mälar links the city to Stockholm. The Swedish love of nature is evident in Uppsala's many parks and open areas.

Gothland

Uppsala is the northern end of a semicircle of cities around Stockholm. The southern end of this arc of cities is Norrköping, a Baltic port in the region of Gothland. Norrköping does not have impressive old buildings such as those that grace Uppsala. But this area, known as Östergötland, or East Gothland, has been inhabited since 6,000 years before the birth of Christ. South of the city, on the Motala

River, are 3,000-year-old rock carvings depicting animals, gods, and a fleet of ships that resemble the longboats of the Goths' descendants, the Vikings.

A city of about 124,000, the seventh largest in Sweden, Norrköping is renowned for its industries, mainly textiles and wood products. Located at the end of the long narrow Bråviken Bay and connected to Stockholm by waterways, highways, and rail lines, it is Sweden's busiest Baltic seaport.

The University of Uppsala, oldest in Sweden, is world famous.

Östergötland, the part of Gothland that lies along the coast south of Stockholm, is more rugged than the plains of Svealand, but the soil is fertile enough for farming. Traditional, bright-red wooden farmhouses stand out against the green hills. This idyllic region is the home of SAAB, Sweden's huge, international automobile manufacturers, and Sweden's flourishing aircraft industry.

Lake Vättern, the last of middle Sweden's great lakes, is 100 miles (161 kilometers) long. It is the second largest lake in Sweden, covering 750 square miles (1,000 square kilometers). The island of Visingsö near the lake's southern end contains the ruins of Visingsborg Castle, one of the strongholds of the 17th- and 18th-century Swedish kings. Several 4,000-year-old tombs on the island are reminders of earlier rulers.

At the lake's southern tip are the cities of Jönköping, Huskvarna, and Gränna. Surrounded by rural villages, Jönköping is a hive of industries making products that range from aircraft to computer software. The historic Swedish Match Company, now a multinational conglomerate, began here 100 years ago with the manufacture of the humble household necessity, the safety match.

Västergötland, or West Gothland, is more rugged than Östergötland because of the "table mountains," abrupt, flat-topped peaks that look a bit like the mesas of the southwestern United States. Bounded on the east by Lake Vättern and on the northwest by Lake

Vänern, West Gothland has many miles of beautiful, forested coastline. Skara, the oldest city in Sweden, is located in the hilly central portion of West Gothland.

Directly west of Gothland is Buhuslan, a strip of wild country along Sweden's west coast that is very much like neighboring Norway. Buhuslan's steep, rocky coastline is pierced by long, narrow ocean inlets that wind for miles between high, heavily forested hills. These inlets are called *fjords*, and their wild beauty attracts tourists from all over the world. Buhuslan is also known for the many prehistoric rock carvings found on its granite cliffs.

Gothenburg

Gothland's largest city, and the second-largest city in Sweden, is Gothenburg, with about 450,000 inhabitants. It lies at the mouth of the Göta River, which carries the waters of Lake Vänern into the

Gothenburg, the principal port of Sweden, shows a mixture of modern and traditional architecture.

Kattegat, the arm of the North Sea below Buhuslan. King Gustavus Adolphus founded Gothenburg in 1619. He wanted a port on Sweden's west coast, because ships from the east could be stopped by ice in the Baltic Sea.

Gothenburg had no natural harbor, so the king brought in Dutch engineers to dredge and widen the rocky mouth of the Göta River. The narrow harbor they created is 14 miles (22 kilometers) long. A forest of steel cranes sways over ships from all over the world. Freighters and tankers are built in the harbor's huge drydocks and giant slipways.

Gothenburg is Sweden's largest port. The majority of Sweden's annual exports, which total over $60 billion, are shipped from here. About a quarter of the country's imports also come through the port, including almost all of the oil Sweden imports annually.

At one end of Gothenberg is Fish Harbor. Every morning fishing boats bring in their North Sea catch to sell at Fish Harbor's market, the largest fish market in Scandinavia. Gothenburg is also the home of Sweden's largest automaker, Volvo, and many banking and financial institutions.

Most Gothenburg residents carry tightly furled umbrellas and the men wear bowler hats, like the businessmen in the financial district of London. In fact, Gothenburg is sometimes called "Little London." During the British and French wars at the beginning of the 19th century, France blockaded the British Isles for several years. Many English and Scottish people moved their businesses to Gothenburg, the nearest neutral port. Some were so successful that they stayed, and many shops and businesses today have such un-Swedish names as Wilson, Chalmers, or McCrae.

The climate is much like London's, too; there is a lot of rain, and the winters are fairly mild. All of Sweden's west coast has a similar climate because of the moisture-laden winds that blow in from the Gulf Stream. Norrland and the central lowlands receive less precipitation, because the rain falls on the Norwegian slopes of the mountains that make up the spine of the Scandinavian peninsula.

Gothenburg is connected to Lake Vänern in the north by the Trollhäte Canal. The Gota Canal links Lake Vänern with Lake Vättern. The canal then continues across the country until it reaches the Baltic Sea south of Norrköping. This series of canals once played an important role in Sweden's economic development. Iron ore and wood were moved along the canals to factories, and steel and sawn lumber were then sent by canal to the country's cities and ports.

North of Norrköping is the Södertälje Canal. A barge trip up the canal takes you through highlands and farmland until the canal ends at Lake Mälar, and the barge docks at Sweden's crown jewel, Stockholm.

An aerial view near the Swedish capital shows the many small islands that form Stockholm's archipelago.

Stockholm and the South

Lake Mälar lies in the center of Svealand. Its long arms stretch in every direction. The barge from Norrköping enters the lake at the end of its eastern arm and passes the red-brick town hall of Stockholm. We go ashore at the foot of Riddarholm Church, the burial place of Sweden's kings. This area is called Gamla Stan, the Old Town. Gamla Stan is actually three islands tied together by bridges and seawalls. It is a cluster of brick and stone houses, royal palaces, and public buildings (such as the Riksdag, or parliament), some of which date from the 13th century.

This is where Stockholm began, about 1,000 years ago, as a fort protecting the entrance to Lake Mälar. The streets behind the public buildings are no more than lanes, and they are as crooked and narrow as they were in the Middle Ages. Today, however, they are lit with electric street lights and crowded with shops and restaurants.

On the eastern side of Gamla Stan is the Saltsjön, an arm of the Baltic Sea. To the north and the south of the Salstjön are the office buildings and apartment blocks, highways and parks, a train station and airline terminal of modern Stockholm. It is a city built around water, with 12 islands and 42 bridges. Almost every street ends in a vista of sea and sky. Stockholm has been called the Venice of the North, but Swedish writer Selma Lagerlöf expressed it better. "Stockholm," she said, "is the city that floats on water."

Stockholm is a very beautiful city, whether it is shining in the intense sunlight of summer or snow-covered and glittering in the long winter night. The population of the metropolitan area (the city and its suburbs) has grown from 100,000 to over 1.5 million in the last century. The municipal and national governments have controlled the city's growth. There are a few tall buildings in downtown Stockholm, but the so-called "grand ensembles"—huge apartment complexes, each with its own set of stores, shops, and essential municipal services—were built to the east and west of the city.

Downtown Stockholm's five districts are knitted together by bridges and green spaces. North of Gamla Stan is Norrmalm, the business district. Here tall buildings surrounded by a web of pedestrian-only streets house Swedish banks and businesses. Nearby is a broad square called the Hotorget. Swedes from all ethnic backgrounds, including immigrants from Turkey, Greece, and Yugoslavia, shop for vegetables and seafood at the open-air market in the Hotorget. Also on the Hotorget is the large concert hall where the Nobel prize ceremony is held every year.

Kungsträdgården (the royal park), is Norrmalm's southern boundary. The park covers the waterfront across the channel from the royal palace in Old Town. A bridge near the park leads to the little island of Skeppsholmen, where a magnificent, white, three-masted sailing ship is docked. This is the *Af Chapman*, a former Swedish navy vessel that now serves as one of the most popular youth hostels in the world.

To the west of Norrmalm is Kungsholmen, the government district. Kungsholmen is a large island in Lake Mälar. Its most remarkable building is the town hall, which is built in an exuberant mixture of architectural styles from many countries and topped with a bright roof of copper shingles.

These tranquil gardens adjoin Drottningholm Palace, home of Sweden's royal family.

East of the business district is Östermalm, an upper-class residential area. In a city where many people live in apartments, the modest brick and frame houses along Östermalm's wide, tree-lined streets represent real luxury. Most foreign embassies are in this neighborhood, as well as many museums and the royal library.

Narvalägen, one of the loveliest of Östermalm's stately boulevards, runs south over a bridge to Djurgården (Deer Garden) Island. Once the royal game preserve, it is now Stockholm's playground. The outdoor museum located on the island houses a vast collection of buildings and artifacts from Sweden's past, including reconstructed peasant homes from Dalarna, a 17th-century chateau (a nobleman's fortified country home or castle), and a Lapp farm.

Another Djurgården museum exhibit is the *Vasa*, which was once the pride of the Swedish imperial navy. The *Vasa* was a 64-gun warship built to order for King Gustavus Adolphus in 1628. On its maiden voyage it suddenly sank in Stockholm Harbor, in full view of the king. The *Vasa* lay in the mud under 36 feet (11 meters) of water until 1961, when it was raised intact. Archaeologists found the ship well preserved. The *Vasa* and its restored contents are now on display at the museum, presenting a fascinating picture of naval life three centuries ago.

Djurgården is the entrance to Stockholm's "Garden of Skerries." Skerries are the many small, usually wooded, islands found along the coasts of Scandinavia. The "Garden of Skerries" extends south into the Baltic, forming an archipelago (group of islands). For centuries these skerries were home to fishermen and farmers. Today, many of Stockholm's inhabitants escape to the skerries for camping vacations.

Stockholm's city planners have left many open spaces in the city and its suburbs. In the winter, ice-fishing shacks dot Lake Mälar, and ice boats race along the quays. The Nacka nature reserve on the outskirts of the city includes a lake and several heavily wooded acres, with miles of groomed and lighted cross-country ski trails.

South of Old Town is Södermalm, another huge island and the last of Stockholm's districts. Södermalm looks somewhat like Montmartre in Paris; brightly painted wooden houses climb the rocky hillsides, and there are sidewalk cafes in tiny squares.

Södermalm is the site of Slussen intersection, one of the busiest traffic spots in Stockholm. Where once a lock allowed boats to pass between Lake Mälar and the Baltic, today multilevel highway interchanges funnel cars and trucks in and out of Stockholm with the least possible disturbance to the neighborhoods below.

Built around water, Stockholm is a city of bridges, a necessity for modern transportation.

Slussen intersection is a symbol of how Sweden tries to tackle the modern world—with ingenuity, respect for the past, and concern for the quality of life of all citizens.

The South

The districts of Skåne and Småland make up Sweden's southern tip. Skåne is a land of fertile plains known as the "breadbasket of Sweden." For much of its history, Skåne was not part of Sweden, but a province of Denmark. To the east of Skåne are the rugged, stony hills of Småland, whose geography and poverty have isolated it from the mainstream of Swedish history.

Geologically, Skåne is more like Denmark than Sweden. Like Denmark's Jutland Peninsula, it is made up of flat sedimentary rock. Over the rock are thick layers of soil that were deposited by the sea when the land was under water and by receding glaciers. Skåne's wide, flat landscape is nourished by the moist Gulf Stream winds; lush crops grow here during the brief but intense Scandinavian summer. The farmhouses are built in the Danish style, half-timbered and whitewashed, and the people share physical and cultural similarities with the Danish.

Skåne is only 2.5 percent of Sweden's total area but it has over 12 percent of the country's population. Approximately 40 percent of the agricultural produce grown in Sweden comes from the rich farmlands of this province.

Malmö is the largest city in Skåne. With 248,000 inhabitants, it is the third-largest city in Sweden. It sits on a sandy, low-lying spit of land just across a narrow channel called the Öresund from Copenhagen, the capital of Denmark. When King Charles X acquired Skåne from Denmark in 1658, Malmö was already a thriving, wealthy city that served as main port of the North Sea herring-fishing fleets. After Malmö's harbor was enlarged at the end of the 18th century, the city's trading importance increased. The harbor, with more than 16 miles (26 kilometers) of quays, is the largest artificial port in Sweden. Goods and people from all over the world—especially Eastern Europe—enter the country at Malmö.

Nineteen miles (30 kilometers) northeast of Malmö is Lund, once the capital of Denmark and first center of Christianity in the pagan land of the Vikings. Founded in 1021 by Denmark's King Canute, Lund was the seat of the first Catholic archbishopric of Scandinavia.

Lund declined in importance after the Reformation, when Sweden rejected Roman Catholicism and made Lutheranism the state religion. In the 19th century, the railroads contributed to Lund's growth as an industrial center. Today it is a very pleasant city with a population of about 97,000 (26,000 of whom are high school and university students).

Northwest of Lund is the port city of Helsingborg. Sitting at the narrowest part of the Öresund, it is only 3 miles (5 kilometers) from the Danish fortress town of Elsinore, the setting for Shakespeare's play *Hamlet*. For many centuries this was the "choke point" for ships entering and leaving the Baltic; whoever controlled Helsingborg controlled trade and supply for most of Scandinavia.

Because of its strategic location, Helsingborg was almost completely destroyed several times, most recently during the disastrous wars of the 17th century. The present port was built in the 19th century, and since then the town has not faced war or destruction because of Sweden's policy of neutrality. Helsingborg's current population is about 110,000. Each year more than 1 million people cross the Öresund by steamer or ferry from Elsinore to Helsingborg. This stream of travelers is still watched over by the massive stone tower known as Karnen (the Keep), a remnant of the mighty medieval fortress that once controlled the Öresund.

Halland faces the Kattegat between Helsingborg and Gothenburg. Ceded to Sweden by Denmark along with Skåne in the 14th century, Halland is a region of heath (treeless, sandy plains swept by sea winds, where heather and other shrubs grow). Before Sweden industrialized, Halland was a poor region whose biggest export was a steady supply of ship captains and hungry sea raiders. Today it is popular vacation country and home to many small industries. Halmstead, the principal city, has 76,000 inhabitants.

Blekinge, the Baltic seacoast northeast of Skåne, is called "the garden of Sweden." Its geography includes all that the Swedes love in their country—fertile valleys, forested hills, and a lovely seacoast. Because it is on the east side of the peninsula, Blekinge gets less rain than the west coast. It has, however, more days of sunshine that anywhere else in Sweden. Blekinge is popular with nature lovers, particularly fishermen. Every spring the king of Sweden opens the fishing season at the tiny village of Morrum, which is famous for its sea trout and salmon.

Småland

The Swedes tell a story about Småland. When God was making Skåne, they say, the Devil sneaked behind him and made Småland. By the time God noticed, it was too late—the Devil had put in the region's harsh, steep, unyielding hills. "All right," said God, "It's too late to change the country, but I'll make the people." And he made the Smålanders. He made them tough, stubborn, independent, and long-lived, so that they could thrive in their land.

This hilly region has large forests, but it does not have great rivers, like Norrland. The soil is thin and rocky, and supports only subsistence farming (farming that grows only enough crops to feed the farmer's family). In the 19th century, harvests were especially poor, and more than one-fourth of Småland's population emigrated. Most came to the United States.

Småland has a long tradition of crafts, such as woodworking, metalworking, knitting, and glassblowing. In earlier days villagers worked together on such crafts in order to earn cash to buy those items they could not grow or make themselves.

This cooperative craft work grew into isolated community factories called *bruks* that set the pattern for Sweden's brand of industrialization. Most Swedish companies employ fewer than 20 people, and employees in even the biggest factories feel they belong to a community of workers.

Glassmaking is Småland's most important industrial craft. The town of Nybro is the home of several glassmaking companies, including Orrefors, whose modern, distinctively Swedish designs are sold in stores worldwide. Some artisans still work in small workshops where they blow glass in the traditional manner, rotating glowing globes of molten glass at the end of long pipes as they sway, cheeks bulging, in the glow of the furnace.

The major city in Småland is the prosperous port of Kalmar. Before Sweden acquired its southern provinces from Denmark, Kalmar was its southernmost port and a very important military and naval base. The huge castle overlooking Kalmar's harbor was begun in the 12th century and enlarged considerably by King Gustav Vasa. Unfortunately, the fortress did not save Kalmar from fire and destruction during wars in 1611 and 1647.

An impressive example of modern Swedish engineering is the great bridge that crosses from Kalmar to the island Öland. The 4-mile (6-kilometer) bridge is the longest in Europe, and it is one of the most exciting and beautiful automobile rides in the world. At midpoint, looking south, one sees no land at all, just a limitless xpanse of glittering sea and blue sky.

Öland and Gorland

Öland is a long, narrow island that is 85 miles (135 kilometers) long and 9.5 miles (15 kilometers) wide at its broadest point. It runs parallel to the coast of Småland, separated from it by a channel called the Kalmarsund. At one point the island is only 4 miles (6 kilometers) from the shore, but the island's terrain and climate are so different that it seems like another country.

Because the island is in the rain shadow of the mainland, there is little precipitation, especially on the east coast. The island is made up of porous limestone rock covered by thin soil, so rain and melted snow drain quickly away. The level center of the island is a steppe, a treeless plain covered with grasses and weeds that can tolerate dry, cool conditions.

Öland's west coast has better soil and more rain, and people have farmed here for centuries. Because farmers must dig deep wells to find water for their fields and homes, windmills are used to pump the well water to the surface. There are more than 600 windmills on the island, making part of it look like Holland.

Öland's many fine beaches make the island popular in the summer. There are also fir forests, a marsh, and the desolate steppe. Hundreds of different migratory birds stop at the bird sanctuary on the southern tip of the island.

Many reminders of the past are found on Öland. The only town, the seaside resort of Borgholm, is dominated by the remains of a 13th-century castle. The stone tomb of a Viking chieftain on the northern tip of the island is shaped like a Viking longship and has been poised on the edge of the Baltic for 12 centuries. Sixteen

The Cathedral of St. Mary in Visby on the island of Gotland, was built in the 13th century.

stone forts built by Iron Age tribes are reminders of even earlier inhabitants.

Northeast of Öland is Sweden's largest island, Gotland. Gotland is also made of limestone, but the island's water supply is better than Öland's. Farms are found throughout Gotland. As on Öland, tourists flock to Gotland during the summer. The greatest tourist attraction is Visby, the island's only town. The older section of Visby around the harbor is surrounded on three sides by a stone wall nearly 2 miles (1.3 kilometers) long. The wall and its 38 towers were built in the 13th and 14th centuries, when Visby, then owned by Germany, was the second largest and most important trading center in northern Europe. Ships from all over the world tied up in its harbor, and goods from China, Russia, and other countries were traded for the furs and minerals of the north. The wall protected the German traders from the resentful Gotlanders and the armies of the Swedes and Danes.

The wall did its job until 1361, when Visby fell to the Danes. The great warehouses and mansions inside the walls are in ruins. Bright red roses grow over the remains of the 16 churches that once flourished there. Outside, green fields lap up against the walls. The turbulence and violence of the past have been replaced by peace and beauty.

Archeologists have found many historic artifacts at excavation sites in Stockholm.

Early History

People have lived in Sweden since the glaciers retreated at the end of the last Ice Age. Stone Age hunter-gatherers wandered north and settled in the warm and fertile plains of Skåne and Gothland. Over the centuries they began to plant and raise crops.

The plows of present-day farmers sometimes turn up prehistoric tools, such as axes, knives, and scrapers, made of chipped stone with polished edges. Polishing the stone edges made the tools sharper and strong. These polished stone tools are from the New Stone Age, and the race that flourished then is called *neolithic* (from the Latin words for "new" and "stone").

The neolithic people who inhabited prehistoric Sweden buried their important dead in earth-covered stone tombs called barrows or tumuli. They also built structures called dolmens. The dolmens were a circle of supporting horizontal stone slabs. No one knows why the dolmens were built. Stonehenge in England has some of the world's most famous dolmens.

The largest barrow in northern Europe is at Barsebäck in Skåne. The wedge-shaped stone tomb is still covered by a huge mound of earth. Inside, a long passage leads to the burial chamber that is 6 feet (2 meters) high and 29 feet (8.5 meters) wide. It is estimated that the stones that make up the roof weigh 20 tons (18 metric tons).

The neolithic people were invaded and overtaken by a new race that had learned how to mine copper and tin and smelt them together to make bronze. With bronze they made stronger tools—and more deadly weapons.

The Boat-Axe People

The Bronze Age, as it is called now, began in Sweden about 2000 B.C. The Aryans, a nomadic race from the steppes of Russia, invaded the Scandinavian peninsula. Aryans also overran southern Europe and India. They brought the Bronze Age culture with them.

The Aryans who drove out or enslaved Scandinavia's former inhabitants are called the Boat-Axe people. The Boat-Axe people, who were named after the shape of their weapons, made up several different tribes. Each tribe settled on its own conquered territory, or *land*. The names of the tribes echo today in the names of Sweden's geographic regions: Svealand, Gothland, and Småland.

The Boat-Axe people also buried their chieftains in barrows. Over the centuries, farmers vandalized many barrows looking for gold, leaving few implements for archaeologists to study. Many rock carvings from the period remain, however, on barrow walls and cliff faces. Archaeologists who have studied these carvings believe that the Boat-Axe people's nomadic way of life changed quickly as they set up villages and began to farm. Fleets of boats carved in the rocks, indicate that seafaring began early.

The carvings show that the Boat-Axe people worshipped several gods, including the sun, which was represented by a disc carried on a ship or a cart. They also worshipped fertility gods. One rock carving shows a vertical pole with a horizontal crosspiece and a horned figure at the top. Three men hold ribbons tied to the pole's crosspiece. The carving is of a maypole, an almost universal, ancient symbol of fertility. Men and women danced around the maypole in the spring, to welcome the sun and to ensure good crops and fertile herds.

The custom survives even today in the rural areas of Sweden. Every Midsummer Eve, when the sun stays in the sky until past midnight, boys and girls dressed in bright folk costumes dance round a maypole, while a huge bonfire burns until dawn. Since Sweden became Christianized, the festival has been called St. John's Day.

Other rock carvings suggest that Sweden's early inhabitants carried out human sacrifice as part of their religion. Human sacrifice continued in Sweden until the Viking Age.

The period from 1500 B.C. to 500 B.C. was stable and prosperous, according to the archaeological evidence. Tools, ornaments, and coins from southern Europe indicate that Sweden's early inhabitants traded furs, amber, and slaves with that part of the world.

Around 500 B.C., Europe's climate turned colder, and a new wave of migration washed across the continent. Nomadic tribes again swept out of the Russian steppes, looking for food and land.

Sweden escaped invasion, but trade with Mediterranean countries ceased. Crops failed, and Sweden's tribes fought over scarce resources. Stone fortresses such as those on Öland Island were built during this time.

The fortresses were not permanent fortifications like the castles built during the Middle Ages. Instead, they provided temporary shelter during attacks. People lived near the fortresses on farms. When an attacking tribe was sighted, people would flee to the fortresses with food, livestock, and weapons.

Iron and Writing

Gradually, the climate improved and the migrations and invasions in northern Europe ended. Trade with the Mediterranean civilizations resumed, bringing to Sweden two great technologies—iron-making and a written language.

Humans first learned to mine and smelt iron in the Middle East; about 1500 B.C. Phoenician traders and other travelers spread the skill among the Mediterranean peoples. The Romans introduced iron making to Swedish tribes about 100 B.C. The Scandinavians also adopted the written alphabet of the Greeks and Romans. Because most of their writing was carved on stone or wood, the letters, called runes, were made entirely with straight strokes.

Rock carvings, such as this one from 2000 B.C., help us understand the level of early Viking civilization.

A museum restoration expert examines an ancient Viking pot.

By 700 A.D., the Germanic tribes of Scandinavia had evolved into the Nordic Vikings. They built strong, seaworthy ships, and learned how to use sails.

The few isolated farms grew in size. Small tribes merged into larger tribes, and their chieftains elected kings to lead them in war or on raids. In Sweden, the Sveas dominated the Goths, but each tribe maintained separate homelands until the 10th century.

It is only from around this time that Denmark, Norway, and Sweden can be thought of as separate countries, even though their boundaries were vague and would change over the years. Sweden consisted only of Svealand, Gothland, and some parts of Norrland.

The Vikings

"From the fury of the Norseman, O Lord, deliver us," prayed 10th century Catholics in northern France. Villagers in England, monks in Scotland, and merchants in Constantinople would have joined in a fervent "amen." Beginning in 800 A.D., the Scandinavians erupted out of their northern isolation and changed the face of Europe.

Each spring, after the crops were planted, Swedish, Norwegian, and Danish men launched their long warships and went roving in search of plunder, land, and trade. These were called Vikings, which comes from an ancient word for bandit camps. Where they headed depended on geography; what they did depended on circumstance. The Norwegians were in the far north and west of Europe, so they sailed west, raiding villages and monasteries in Ireland and Scotland. They also discovered and colonized Iceland, Greenland, and the coast of North America. The Danes moved into the English Channel and raided the coasts of England and France. In France, they colonized the province of Normandy. In England, they controlled the entire northern half of the country and collected an annual bribe, called the Danegeld, for sparing the southern half of England from raids.

Because Swedish ports were on the eastern side of the peninsula, the Swedish Vikings sailed to the east across the Baltic Sea. Landing on the coast of what is now Lithuania in the Soviet Union, they used logs to roll their longships to the Dnieper and Volga rivers. The Vikings then sailed down to the Black Sea and the Caspian Sea, and to the rich cities of Constantinople and northern Persia.

Along the way, they followed the traditional Viking policy of taking from the weak and trading with the strong. The Vikings tried to sack Constantinople twice, in 860 and 941, but failed both times. They then colonized and fortified the river route to the Middle East, and required annual bribes from the Slavs living in the surrounding countryside.

The Slavs called the Swedish Vikings "Rus," and the land that they controlled came to be called Russia. One Swedish Viking chieftain became King of Kiev. However, the Vikings lost their hold over the Slavs and their trade route was taken over by the Tartars, who

(continued on page 65)

SCENES OF SWEDEN

▼ The streets leading from Gamla Stan's central square, Stortorget, are alive with the fashion, food, and culture of the day.

◄ This view of Gothenburg at night shows Carl Milles's famous sculpture.
▼ The Old Town, known as Gamla Stan, is the original heart of Stockholm.

◄ Each year the City Hall in Stockholm is witness to a spectacular event called "Sailboat Day."

▼ The Golden Coast extends 220 miles (352 kilometers) up to the Norwegian border, inviting swimming, recreation, and relaxation.

◄ Swedish crystal and glassware are world famous for their clarity and clean design.

▲ *Malmö, Sweden's third-largest city, lies on its southern coast, a short distance from Copenhagen, Denmark.*

◄ *The west coast archipelago near Gothenburg has one of the world's finest fishing grounds.*

➤ *A young hiker rests by a stream in the beautiful Swedish countryside.*

◄ *These compact wooded areas abound with kilns for glassworks.*

↖ *A group of young Swedes in Dalarna celebrate Midsummer in traditional local costumes.*

▲ A portion of the Lapps continue to live a nomadic life with their herds of reindeer.

◄ St. Lucia Day is celebrated several weeks before Christmas.

invaded Russia in the 11th century. The Swedish Vikings left few traces of their rule over Russia. The most conspicuous relics are the thousands of barrows along the Dnieper and Volga rivers.

In 922, an Arab trader witnessed the funeral of a Viking chieftain by the Volga River. The trader said the Vikings attending the funeral were tall, well built, and red-haired. He noted that they were very dirty. Each man was armed with an axe, knife, helmet, shield, and two-handed sword. The Viking women wore metal caps over their breasts.

The dead chief was placed fully clothed in a beached ship stocked with weapons, clothing, food, and drink. A drugged slave woman was raped by the senior Viking men, and then strangled and placed on the ship. The ship was then burned, and the ashes covered with a mound of dirt and a rune stone.

Christians considered the Vikings pagans, because they worshipped many gods. The chief Viking god was Odin, known as father of the gods and the god of the hanged. Every nine years, nine males from each animal species—including humans—were hanged on a sacred yew tree in Uppsala to honor Odin. Thor was the god of war, whose hammer gave forth lightning, and Frey was the god of fertility. The Vikings believed that Frey was the father of the Yngling Dynasty, the half-legendary first kings of Sweden.

At the same time that they were profoundly influencing the rest of the world, the Vikings were being changed by their raids and conquests. The yearly expeditions brought back not only silk, spices, and silver, but also new ideas.

One idea was the concept of country, or nation. Until the 11th century, if you asked a Viking where he was from, he would have given you the name of his tribe or the leader of his horde (the temporary alliance of tribes under one leader). Gradually, however, as tribes cooperated on raids and began to share their resources and ideas, the idea took root that a country and its people have an identity that goes beyond tribe and horde. Their contact with peoples in distant parts of the world helped the Vikings recognize that their own culture was distinct and included all tribes.

Another powerful idea that returned with the Vikings was Christianity. The first Christian missionary came to Sweden in 829, and surprisingly, he was not martyred. In fact, the Vikings were tolerant

The Vikings, ancestors of today's Swedes, terrorized Europe during the Dark Ages.

of other religions. They would allow the worship of a new god, but they reacted violently when Christian priests tried to banish the old gods.

There were many Christian martyrs to the new religion before it was finally accepted, and there were many battles before Sweden became a country. In the year 1006, the ideas of Christianity and country reached a culmination with the baptism of Olof Skötkonung, the first recorded king of a Swedish state larger than a horde.

Danish King Christian II was defeated by the Swedes in a naval battle in Stockholm's harbor.

Building the Swedish State

Because of its remoteness, Sweden was one of the last European countries to accept Christianity. For two centuries after King Olof Skötkonung's baptism, pagans and Christians fought for control of the kingdom. One of the early Christian kings was called Eric. Eric invaded what is now Finland, baptized the pagans there, and annexed the land to Sweden. After his death in battle in 1160, he was canonized and became the patron saint of Sweden.

The Viking Age came to a gradual end around 1160. The countries of Western Europe had developed stronger governments and were no longer easy targets for the Vikings. Viking settlements along the Dnieper and Volga rivers in Russia were overrun during the Tartar invasion in the 11th century.

The structure of society also changed as the Viking Age came to an end. Viking society had only three classes: *thralls*, *karls*, and *jarls*. The lowest social class was the *thralls*. These people were similar to the serf class in Western Europe; they were treated as a rich person's property and forced to farm the lands of the rich. Next on the social scale were the *karls*, free men who owned their own farms. The third and highest class was the warrior-chiefs, or *jarls*. Jarls protected the karls and led them in battle.

Civil war between rival kings and territorial war with Denmark led to the development of a new class. Farmers and minor chiefs had turned to the stronger jarls for leadership and security toward the end of the Viking Age. In return for leadership and protection, however, men had to swear an oath of loyalty and obedience to the jarl or lord. These lords eventually obtained large tracts of land, or estates, from the farmers and minor chiefs. The people who lived on an estate owed their livelihood to the lord and his heirs. These nobles in turn were supposed to support and obey the king.

This social structure is called feudalism. Feudalism developed in Europe in the 5th century, but did not become established in Sweden until the end of the 12th century. However, the Swedish version was different. The kingship remained elective, instead of hereditary, until the 15th century. And Swedish farmers living on the estates kept more individual rights than the feudal farmers in the rest of Europe.

Several noble families became powerful in Sweden. One was the Folkung family, whose leader, Birger Jarl, became the trusted confidant of King Erik Eriksson, descendant of St. Erik. Birger Jarl actually ruled the country. He stopped the civil war by curbing the power of the nobles and by increasing the strength of the central government. Birger's son Valdemar was elected king in 1250 after King Eric's death. Valdemar's brother Magnus overthrew him in 1278.

Birger Jarl and his sons consolidated Sweden's conquests in Finland and began building Stockholm. Trade and business improved with the political stability. In 1347, the royal charter for developing copper mines at Falun was granted. New towns were built, including Visby, the trading center on the island of Gotland.

Visby was founded by Germans who were members of the Hanseatic League, an association of independent, mostly German, cities within the Holy Roman Empire that covered most of northern Europe. The league cities controlled all the trade in the Baltic regions. Birger Jarl's sons increased trade with the league in metals, furs, and wood.

Unfortunately, Magnus's descendants were not as capable as he, and the Folkung family was overthrown in 1318. Noble families again fought among themselves for control of the country, while Danish kings and German princes threatened Swedish control of lands in Finland and Norway.

In 1397, Sweden, Norway, and Denmark united to defend themselves from German domination. In a treaty signed at the Swedish fortress of Kalmar, the three nations made Queen Margaret of Denmark the ruler of all three countries. Called the Kalmar Union, this confederation kept the Hanseatic League from dominating Scandinavia.

In 1412, Margaret was succeeded as head of the union by her son, Erik. Erik's war on the German armies required large taxes that were resented by the Swedes. In 1434, Swedish peasants and miners revolted against their absent king. They were led by a knight called Engelbreckt Engelbrecktsson.

The Peasants' Rebellion

The peasants' rebellion led to the establishment of Sweden's first Riksdag, or parliament. The Riksdag's members were representatives of the four "estates" or classes: the nobility, the clergy, the burghers (middle-class craftsmen and traders), and the peasants. The Riksdag was the first legislative body in Europe to allow peasant representatives, and the first to meet independently of the king. Because of this, the Swedish Riksdag is considered the first democratic parliament.

Erik was deposed in 1436 and the Swedish monarchy again became the focus of fierce struggle among the nobility. The Kalmar Union persisted, although it became dominated by the Danes.

The kings of Denmark considered Sweden and Norway possessions of the Danish throne. In 1520, Denmark's King Christian II invaded Sweden. After a four-month siege, he captured Stockholm, where he was crowned king of Sweden. In spite of the amnesty he announced when the Swedes surrendered, he executed 80 of Sweden's leading nobles in the public square of Stockholm. Christian then left the bodies in the square for several days before they were burned in a ditch outside the city walls.

This act of savage treachery became known as the Stockholm Bloodbath. It led to Christian's downfall, the end of the Kalmar Union, and Sweden's establishment as a modern nation. One of the few Swedish noblemen to escape the Stockholm Bloodbath was Gustav Vasa. He fled to Dalarna, raised an army, and drove the Danes out of Sweden. In 1532, he was elected king.

Swedes consider Gustav Vasa the father of their country. He reorganized the state and brought Sweden out of the Middle Ages and into the Renaissance. As one of his reforms, he confiscated church lands. The Catholic church owned more than one-fifth of all the land in Sweden. The church's holdings included the best farmland; moreover, the church did not pay taxes on the land it owned. The peasants resented the church's control of the land. Gustav took advantage of the peasants' resentment and the spreading belief in Lutheranism to seize the Catholic church land for the state.

With a wealthy and strong central government, thriving mines and mills, and a disciplined army, Sweden began to expand under Gustav and his heirs. Wars were fought with Denmark, Poland, the Holy Roman Empire, and Russia for control of the Baltic Sea. The Baltic was an important trading area for the European nations.

Gustav Vasa's grandson, Gustavus II Adolphus, ruled from 1611 to 1632 and founded the city of Gothenburg in 1619. He led the best-trained and best-equipped army in Europe on the Protestant side during the Thirty Years' War, the last major religious war of

Gustav Vasa, who brought Sweden out of the Middle Ages, is considered the father of his country.

the Reformation. Adolphus won control of important Baltic ports. He was killed in battle, and his daughter, Christina, succeeded him and ruled from 1644 to 1655. Under Christina's rule, Sweden gained Gotland from Denmark and Jämtland from Norway.

For the first time since the Viking Age, Swedes began to look at the world beyond the Baltic for adventure, land, and treasure. In 1638, Sweden founded the colony of Nya Sverige (New Sweden) in the Delaware River Valley of North America. (The Dutch took over the colony in 1655.) In the 1650s, Swedish traders established a trading post in Africa's Gold Coast. The Dutch also took over this colony.

In 1657, Denmark invaded Sweden while Christina's successor, Charles X Gustav (who ruled from 1654 to 1660), was at war in Poland. Charles retaliated and invaded Denmark. He led a march

Gustavus Adolphus II was king of Sweden and one of its leading generals during the Thirty Years War.

across the frozen Öresund and surprised the Danish forces. The Swedes almost captured Copenhagen in the invasion. To save the capital city, Denmark was forced to give Sweden the provinces of Skåne, Halland, and Blekinge in 1658. Sweden finally controlled all the land to the Baltic coast.

The Invasion of Russia

Sweden's greatest military glory came under the reign of Charles X's grandson, Charles XII, who ruled from 1697 to 1718. Russia's czar, Peter the Great, who wanted to control the Baltic, joined with Denmark and Poland in a war against Sweden. Charles defeated Peter's army at Narva in 1700. Charles then invaded Poland. His armies were usually outnumbered, but their discipline and courage led to victory after victory.

Meanwhile, Peter the Great built a new port called St. Petersburg on the Gulf of Finland. Peter hoped to build a fleet of warships to challenge Swedish domination of the Baltic. In response to the Russian threat, Charles invaded Russia in 1707. The invasion followed a tragic pattern that would be repeated in 1812 by Napoleon and in 1941 by Hitler.

Charles' superb troops easily defeated the larger Russian army. As the Russians retreated before Charles' army, they burned crops and fields behind them. Charles chased the Russian army deep into Russia until the enemy's "scorched earth" tactics left Charles far from his base, with no food for his army.

Instead of retreating to the Baltic, Charles headed south to the fertile Ukraine to spend the winter. In 1708, he was defeated by Peter the Great's army at the Battle of Poltava. Charles and his few thousand remaining soldiers made a daring escape after the battle southward into Turkey.

The tide had turned against Sweden. Charles fought for 10 more years, draining his country of men and money. By 1718, when Charles was killed during an invasion of Norway, Sweden had lost all of it lands beyond the Baltic except Finland. In spite of the constant warring and the harshness of his rule, the people of Sweden honored Charles as a symbol of Swedish courage and honor.

Charles was the last absolute monarch freely followed by the Swedes. They no longer wanted kings whose power was unchecked by the church, the nobles, or the parliament. Charles's death signaled the beginning of the "Age of Liberty." Because Charles left no heirs, the Riksdag elected new rulers—his sister Ulrika and her husband Fredrik—on condition that they accept a new constitution.

The new constitution gave most power to a council selected by the Riksdag. Under the council's rule, Sweden turned again to farming, fishing, and trade. The market for high-quality Swedish iron grew steadily, bolstering the Swedish economy. Sweden remained the major sea power in the Baltic.

Unfortunately, peace did not last long. Political freedom gave rise to two parties, nicknamed the Caps and the Hats. The Caps, associated with older nobles and farmers, favored peace and free trade. The Hats were associated with younger nobles, burghers who wanted tariffs (taxes on imports), and the army. They yearned for revenge on Russia and a restoration of Swedish glory.

The mid-18th century was a period of much confusion, including a disastrous attempt to invade Russia and an attempt to overthrow the Riksdag. Finally, in 1772, King Gustav III staged a coup. He disbanded the council and restored the absolute monarchy.

The end of the Age of Liberty was greeted with relief by many Swedes, because political instability and bad luck had led to unemployment, crop failures, and social unrest. Gustav was a sound administrator as well as a good politician. He reformed the economy and calmed the people. He made some social reforms, such as repealing the death penalty for witchcraft, allowing freedom of worship for foreigners, and lifting some trade restrictions. He was a patron of the arts and of philosophy, and for a time Stockholm glittered with the gaiety and sophistication of Paris. Nevertheless, Gustav was a stern ruler whose reign had elements of tyranny.

Gustav's power enraged and frightened the nobles. In March 1792, an army officer who was a nobleman shot and fatally wounded the king at a masked ball at the Stockholm Opera House. The assassin and the other conspirators had hoped to restore power to the nobility, but the people of Sweden reacted angrily to the killing. The king's son, Gustav IV Adolf, succeeded Gustav to the throne.

The restored monarchy survived Gustav's assassination, but not the Napoleonic wars. Sweden sided with Britain, Russia, and Austria against France's Napoleon. When Russia changed sides, Sweden faced the combined armies of Russia, Denmark, and France. The British fleet saved Sweden from invasion, but Swedish troops were driven out of Finland by Russia.

Humiliated and threatened with total defeat, the Swedish army revolted and deposed Gustav in 1809. A new constitution was written in 1809 that gave political power back to the Riksdag, and a weak king, Charles XIII, was crowned. Napoleon forced Sweden to sue Russia for peace. The Russians demanded—and received—all of Finland and part of Norrland.

The Reign of Jean Bernadotte

Charles ruled for less than a decade. Impressed with Napoleon's strength, the Swedish government offered the Swedish monarchy to one of Napoleon's marshals, Jean Bernadotte. Bernadotte accepted. Taking the name Charles John, he became crown prince in 1810 and king in 1818. The Bernadotte dynasty continues today—the present king, Charles XVI Gustav, is a direct descendant.

Bernadotte surprised many Swedes by making Sweden an ally of Russia. He thus accepted the loss of Finland, but he was able to seize Norway from Denmark in 1814 in the Peace of Kiel. The union between Sweden and Norway lasted until Norway became independent in 1905.

Since the Peace of Kiel, Sweden has not been at war. The country has followed a policy of "armed neutrality," keeping a standing army and a highly developed civil defense system.

The French revolutionary who became Sweden's king turned out to be a very conservative ruler. The new constitution of 1809 provided for a balance of power between the king and the Riksdag, but Charles John resisted reforms that would limit his power. He suppressed newspapers and arrested his critics, because he believed opposition to his policies was disruptive and irresponsible.

Many Swedes agreed with the king's policies. They preferred a strong ruler who could return Sweden to its days of glory. But there was a strong movement advocating reform of the government and of the social structure. This movement found expression in newspapers, books, and even street riots. Charles John was forced to allow some changes. When he died in 1844, many important reforms were carried out under his son Oscar I, who ruled from 1844 to 1859.

Oscar allowed free trade and free enterprise. The Riksdag was made bicameral (that is, having two houses). This resulted in more equal representation for the various classes, although the right to vote was still limited to men who owned property. The reform with the greatest effect on Swedish life was the Education Bill of 1842, which established compulsory education for all children.

The Great Depression and Migration

Around 1860, an economic crisis overwhelmed the reform movement. Thanks to improved public health, better food supply, and no war ("peace, vaccination, and potatoes," as one historian put it), the population had increased from about 2.3 million before 1800 to almost 4 million. Swedish iron mills lost customers to British mills, which had developed new processes for making iron more cheaply. The logging industry was depressed. There were no jobs, and the countryside was full of people with no money, no land, and no hope.

The suicide of Sweden's "Match King" Ivar Kreuger precipitated a worldwide financial crisis.

Many of the unemployed and poor decided to emigrate to America, where early Swedish immigrants in the 1830s had found forests, lakes, and—most important—work and the prospect of owning land. By 1850, 15,000 Swedes lived in the United States.

Poor harvests in Sweden in the 1860s swelled the number who left. Between 1868 and 1870, over 80,000 Swedes sailed to the United States. Even though Swedish industry made a strong recovery from the slump of the 1860s, the exodus of poor, land-hungry country people continued. The peak occurred in the 1880s, when an estimated 325,000 went to the United States. Increasing numbers of the new immigrants settled in large cities such as Chicago, which had 50,000 Swedish inhabitants in 1900.

By 1880 the economy had changed greatly. Swedish inventors and investors helped the iron industry produce more and better iron. The development of cheap paper from wood pulp and the invention of the steam sawmill boosted the Swedish logging industry. But these economic changes were too late to halt the flood of emigration.

The Swedes who remained continued remaking Sweden's economy and society. The Industrial Revolution, like Christianity, feudalism, and the Reformation, came to Sweden later than it did to other societies. In many ways, Sweden was ready for it. The rekindled economy had given Sweden a well-fed, well-educated working class. The many lakes and rivers provided cheap water power and transportation. The small craftwork factories of Småland and the iron mills of the central region provided the basis for a vital, powerful industrial economy.

Industrialization also brought problems. Many factories and shops were dirty and dangerous. Employers paid workers low wages. There was labor unrest. At first strikes were ruthlessly ended with troops, but after the Swedish labor movement united under the leadership of the socialist August Palm, the government was forced to grant the workers greater freedom of expression. In 1908, the movement staged a general strike in which more than 300,000 men took part.

Creating a Democratic Government

As the 20th century began, Swedish society and politics became increasingly open and democratic. The monarchy was changing into a constitutional monarchy—that is, one where decisions are made in the king's name by the elected representatives of the people. Three political parties had emerged with the increase in political freedom—Conservative, Liberal, and Social Democratic. They differed on issues such as defense, the power of the king, and, most important, economics. All agreed, however, on granting more rights to the workers.

Sweden's policy of armed neutrality kept Sweden out of World War I. At the close of the war, Sweden was politically and economically stable. The Communist revolution in Russia in 1917, however, alarmed moderates and stirred up radicals in Sweden. The radicals inspired a sweeping reform of the political system: all citizens now had the right to vote, and women could be elected to political positions.

After World War I, Swedish industry began to sell its products on the world market, and Swedish financiers began to invest in businesses throughout the world. Occasional economic slowdowns and business failures, however, led to unemployment, strikes, and calls for socialist (government ownership of land and industry) reform or even communist revolution.

The worldwide economic depression in the 1930s put hundreds of thousands of Swedes out of work and into government-sponsored public work projects. The collapse during the depression of the giant international Swedish Match Company was a turning point in Swedish politics. The head of the company, Ivar Kreuger, had been a hero to Swedes. When his kingdom crashed, million of dollars vanished overnight. Kreuger committed suicide. Investigation showed that the

The horse-drawn carriage containing the remains of King Gustaf VI Adolf passes mourners in this 1973 funeral procession.

company's vast holdings were largely fake and that government officials had been bribed.

The furor over the collapse and the despair of the Swedish people led to a landslide election victory in 1932 for the socialist Social Democratic Labor party. The Social Democrats transformed Sweden into a welfare state—one in which the government ensures the health, education, and economic security of every citizen.

Sweden never became a true socialist state, however. The Social Democrats, who have been in power almost continually since 1932, have followed what they describe as "the middle way" between socialism and capitalism.

World War II, which broke out in 1939, put Sweden's policy of neutrality to an extreme test. Sweden immediately declared its neutrality when the war began. It refused to allow British and other troops to cross through Sweden to attack the German army. Later in the war, after the Germans occupied Denmark and Norway, they threatened to attack Sweden if it did not allow Germany to transport troops from Norway through Sweden. Sweden complied. In spite of this compromise, neutrality allowed Sweden to escape devastating damage during the war. Unlike most other European nations, the country emerged from the war with a functioning economy and a stable government.

At the end of the war, Sweden became one of the first members of United Nations. The Social Democratic party remained in power. Between 1946 and 1950, the Social Democrats launched many of the health, education, and other social welfare programs that Swedes take for granted today.

In 1955, a committee reviewing the 1809 constitution recommended the replacement of the bicameral Riksdag with a unicameral, or one-house, parliament. The king's powers were also reduced to a purely ceremonial role.

Sweden's economy performed energetically in the postwar years. A few years of unemployment and high inflation in the 1950s caused much concern among Swedes, but industrial and forestry production soon grew again and the standard of living did not drop. The oil crisis of the 1970s, however, led to poor economic performance and the end of 44 years of Social Democratic rule.

In the 1976 elections, the Social Democrats were replaced by the Centre party. The Centre party tried to end Sweden's nuclear power program at a time when Sweden was feeling the effects of dependence on foreign oil. The proposal led to intense and heated debates. In 1980, a referendum was held in which voters agreed to phase out all nuclear power plants and use alternative energy sources.

The Social Democrats under Olof Palme, who became prime minister, were returned to power in 1982. Palme introduced new taxes on corporate profits. This move led to widespread demonstrations by employer groups. Palme and the Social Democrats won the 1985 general election, but Palme's tenure as prime minister was cut short when an unknown assassin shot him to death in a Stockholm street.

The killing shocked Swedes. Palme and other Swedish leaders had never had bodyguards. Political violence had been unknown until his killing.

Ingvar Carlsson took over Palme's position. Carlsson continued the Social Democrats' social welfare program, but there was a growing debate over the high taxes required to pay for the extensive social services Swedes receive. By 1987, taxes were absorbing 55 percent of Sweden's gross domestic product (GDP). The average for other industrialized countries was 40 percent. In the elections held in 1991, the Social Democrats once again lost power. They were replaced by a conservative coalition that promised to reduce taxes and cut back on the welfare state.

The Social Democrats returned to power in 1994. They failed to win a majority of the seats in the Riksdag, but they formed a minority government with the support of the Centre Party.

In November 1994, in a nationwide referendum, the Swedish people voted to join the European Union (EU). In January 1995 Sweden became a full member of the EU. Ingvar Carlsson retired in 1996, as he had promised he would, and the finance minister, Göran Persson, replaced him as prime minister.

In 1947, King Gustaf V, 89 years old, opened the Riksdag (Swedish parliament) for the forty-first time in his reign.

Government and Economy

Sweden is a constitutional monarchy. The king or queen is considered the head of state, but true political power rests with the people. The monarch now serves only as a figurehead, a far cry from the absolute power wielded by former kings and queens.

The 349-member parliament, or Riksdag, formulates the policies and makes the laws that guide Sweden. All citizens over the age of 18 elect representatives to the parliament. The members of the Riksdag elect from among themselves a cabinet, which acts as the government. The cabinet members, who are called "ministers," run the government and administer the laws passed by the Riksdag. The prime minister is the effective head of state.

Sweden's parliament is the oldest in Europe. In 1971, it was changed from a bicameral to a unicameral parliament. Bicameral parliaments, such as the British parliament, have two houses. One house (usually called the lower house) is made up of elected members. Members of the other (upper) house are appointed. In Sweden's unicameral parliament, all the representatives are elected. The Riksdag's 349 members are elected to three-year terms by the people in the towns or counties that they represent. More than 90 percent of eligible voters vote in the elections.

Women make up over 40 percent of the Riksdag's membership.

Sweden has several major political parties. In the 1994 elections, for example, seven parties gained seats in the Riksdag. Because one party rarely wins a majority of seats, members from two or more parties must agree to vote together in order to elect a cabinet and prime minister. Such arrangements are called coalition governments. The Social Democratic party has ruled Sweden, either by itself or in coalition, almost continually since 1932.

The cabinet usually has 18 to 20 ministers, each in charge of his or her own ministry, such as justice, defense, or education. The ministers make budgets and work for the passage of laws that concern their ministry. The civil service carries out the business of government, such as collecting taxes and administering welfare programs. Civil servants work for the government, but not for the elected representatives. They are hired and promoted on merit, and their jobs and careers are independent of party politics.

Local elections are also held every three years for county and municipal offices. The county governments are mainly concerned with the administration of the health-care system. The municipal governments administer the schools, city planning, housing, and social welfare issues, such as child, youth, and old-age care.

A unique Swedish institution that originated in the 19th century is the parliamentary ombudsman. Ombudsmen look out for the rights of individual citizens. They monitor how authorities use their power, and they prosecute officials and civil servants who break the law. Three of the officers serve on the national level and more serve at the local levels of government.

The Swedish judicial system has three levels of courts: local courts, appeal courts, and the Supreme Court. Most civil and criminal trials are conducted in local courts. A local court consists of a judge and a panel of lay assessors (nonlawyers appointed by the municipal councils). Juries of regular citizens are used only in cases of freedom of the press.

Local courts are very careful to match sentences to the accused's crime. The emphasis in criminal cases is on rehabilitation instead of punishment. Many sentences are suspended, and the convicted person is put on probation. There is no capital punishment. Most prisons are open and unwalled, and look like college dormitories. Inmates work at jobs in the community, and they are routinely given furloughs to visit their homes.

The cornerstone of Sweden's foreign policy is armed neutrality: this means that Sweden will avoid involvement in all wars unless it is directly attacked. Since 1814 it has successfully followed this policy, which has required Sweden not to join in any military alliances. After World War II, Sweden eagerly joined the United Nations but refused to join the North Atlantic Treaty Organization (NATO). Following the same reasoning, Sweden was long reluctant to join the European Economic Community and its successor, the European

This Swedish tank is symbolic of Sweden's strong military force, designed to ensure her neutrality.

Union (EU). Sweden feared that membership in such an organization would limit its ability to remain neutral in conflicts involving other European countries. In 1995, however, with economic cooperation becoming even more important and direct military threats seeming at least temporarily remote, Sweden did become a full-fledged member of the EU.

Even during its years outside the EU, Sweden cooperated closely with its Scandinavian neighbors in all but military matters. For example, long before the EU nations began reducing trade and travel barriers, Sweden joined with Denmark, Finland, and Iceland to issue a joint passport. A citizen of any of those countries was able to work, live, and receive full social benefits in any other one.

A strong defense has always been required to maintain Sweden's policy of armed neutrality. The country still spends about 2.5 percent of its gross domestic product (GDP) on its armed forces each year—a substantial proportion of its total government expenditures. The armed services have about 640,000 men and women in their ranks. Every man must serve about 10 months in the military. In addition, he must attend refresher courses and perform reserve duty. Fully mobilized, the armed forces of Sweden would number about 729,000.

A sophisticated system of air and sea defense protects Sweden. Ground and sea radar sweep the skies constantly. Submarine chasers search the Baltic for foreign submarines that might come too close to the coast. In the event of an attack, Swedish-built jet fighter-bombers would blast out of mountain caves and island airfields to defend the country.

Civil defense is a high priority. Every town and city has extensive underground bomb shelters that can protect the population for weeks. Air raid drills are held frequently. The shelters, which are carved out of the granite on which most of Sweden rests, are also used as community centers, where arts and craft classes, dances, and town meetings are held.

Sweden has played a leading role in the United Nations. One of the most effective and respected world leaders since World War II was a Swede, Dag Hammarskjöld, the second secretary general of the United Nations. Swedish troops have served in United Nations peacekeeping expeditions in Asia, Africa, and the Middle East. Because it is a neutral country, Sweden criticized both the United States and the Soviet Union during the Cold War.

For years, Sweden has been an active participant in international nuclear disarmament efforts, and it continues to work hard to ban nuclear test explosions and outlaw the development, storage, and use of chemical weapons.

Sweden has an extensive foreign aid program. Since 1977, it has spent 1 percent of its gross domestic product (GDP) on aid to less developed countries, such as Tanzania, Vietnam, and India.

Economic Policy

Sweden's economic policy has been the most successful example of the "middle way." Swedish leaders have tried to find the best compromise between unregulated capitalism and state ownership. Their goal has been to protect the worker without limiting individual freedom.

Sweden has a mixed economy, which means that some businesses are owned by the state. The railroads, telecommunications network, and some of the mining and energy industries are state owned. The state also controls the national health-care, educational, and social welfare systems. The rest of the economy is privately owned or owned and operated by cooperative societies.

Sweden also has a planned economy. Although most businesses are in private hands and operate for profit, the state plays a large role in how they operate by regulating business practices and providing funding to certain industries.

Since the depression, one of the government's most important goals has been to keep unemployment low. In general, the government has been successful. Unemployment in the 1980s averaged under 3 percent. In the mid-1990s, as in much of the rest of Europe, it reached higher levels, around 8 percent.

To keep people employed, the Swedish government uses tax policies, government loans, and other measures to encourage new industries. The government also steps in to help older industries hurt by economic changes. For example, the oil shortages in the 1970s created a worldwide slump in trade that hit the Swedish shipbuilding and steel industries hard. In order to keep the shipyards and mills open, the government provided subsidies and tax breaks.

The arrow indicates the site of the April 1986 accident at the Chernobyl nuclear power plant that spread radioactive materials over Sweden and much of the rest of Europe.

Finally, the government nationalized many of the shipbuilding and steel companies by purchasing them from their stockholders. Nationalizing the businesses means that the government actually owns and operates the business.

Energy is one of the most pressing economic questions in Sweden. Sweden has one of the highest rates of energy consumption in the world. Energy consumption is high because of the cold climate, the heavy energy demands of industry, and the very high standard of living. In Europe, only the Swiss own more cars, televisions, stereos, and other energy-using devices. Sweden's annual energy requirement is the equivalent of over 5.3 tons (4.8 metric tons) of oil per person—and Sweden produces no oil.

The superb system of dams and hydroelectric stations provides 45 percent of Sweden's total electricity supply. However, demand for electricity will grow, and hydropower will not be able to supply the difference.

By the mid-1990s, nuclear power supplied over 40 percent of the nation's electricity. But in accordance with a national referendum in 1980, the government planned to phase out Sweden's nuclear power plants by the early 21st century. The Swedish resolve to do without nuclear power was reinforced by the 1986 disaster at the Soviet Union's Chernobyl nuclear plant.

Like other developed countries, Sweden is dependent on foreign oil. The oil shortages of the 1970s showed how dependent Sweden is on foreign oil. For the first time since World War II, the rate of economic growth slowed and unemployment rose.

This plant extracts uranium ore. Swedish deposits are among the world's most valuable.

Government and industry have worked together to lower energy consumption through education, better engineering, and recycling. Sweden spends more of its GDP on researching and developing new technologies than almost any other country in the world. Much of that effort is now directed toward developing alternative sources of energy, such as solar power, wind power, and biomass (plant and animal waste used as a heat source).

In general, Sweden's planned economy responds well to such challenges, despite fears from some critics that government regulations keep businesses from responding efficiently to market forces.

The 1970s and 1980s saw a switch in the economy. Once Sweden sold raw materials to the world; it now sells finished goods. Forestry, iron and steel production, and ship-building are no longer the most powerful parts of the economy. Instead, engineering drives the economy. Products of the engineering industry account for 50 percent of annual exports, which by the mid-1990s surpassed $60 billion. Exports include precision ball bearings, milking machines, food-processing equipment, automatic lighthouses, luxury automobiles such as the Volvo and the Saab, chemicals, pharmaceuticals, and military and civilian aircraft.

Agriculture employs only about 2 percent of the population, but is very productive. To keep Sweden self-sufficient in basic foods, the government regulates agricultural production and guarantees prices to farmers.

Housing, Transportation, Taxes

Sweden's thriving construction industry is an area of the economy where government planning is considered crucial. From the turn of the century until the late 1970s, Sweden suffered from a severe housing shortage caused by the massive shift in population from the country to the city. Whole families were forced to live in one room and several families had to share one apartment. Families stayed on waiting lists for Stockholm apartments for years.

Gradually, city planners, municipal authorities, and financial institutions responded. The results are the large apartment complexes that encircle most towns and cities. Now almost every Swede has an apartment. Although few people own their own houses, a very high percentage of citizens own or share vacation cottages or cabins. Most Swedish apartments are fairly new and are packed with labor-saving and home-entertainment devices.

Swedes own 3.6 million cars. With one car for every 2.4 people, Sweden has one of the highest concentrations of automobiles in the world. Government control of the automobile industry is thorough. Swedish cars must meet the highest standards for energy conservation and exhaust emissions. The highway system is extensive and well maintained. Sweden has very strictly enforced laws against drunk driving. The laws against drunk driving are supported by the public.

Most of the 4.5 million employed Swedes, including civil servants, belong to a union or trade organization. Pay and working conditions are among the best in the world. The high pay and good working conditions result from collective bargaining agreements between unions and employers. The government does not get involved in labor and employer negotiations unless it is invited by both parties to settle conflicts when negotiations fail.

The government does not set workers' wages, but the government's tax policy affects a worker's standard of living and spending habits. There is a steeply progressive income tax: the more one makes, the higher one's tax rate. The government believes progres-

An average Swedish family is surrounded by the amount of food they consume annually.

sive taxes have an egalitarian effect (that is, income levels become equal). Critics say that the tax system punishes those who succeed and stifles ambition. The government also levies a value-added tax (VAT) on goods and services. Such a tax is like a sales tax, but it is applied at every stage in the production of a good. Each step in the manufacture of a product is determined to add a certain amount of value to the product. Expensive consumer goods such as cameras are taxed the most.

Workers pay on average about 38 percent of their income in state and local taxes. This is an extremely high tax rate, but it pays for a complete range of health and education services.

Health and Education

By the mid-1990s, the Swedish government's annual spending on health and social services averaged $8,500 (65,800 kroner) per citizen. The government's policy is to ensure citizens a guaranteed income and complete health care throughout their lives.

Every citizen is required to contribute to a health insurance plan. Fees for medical care are very low. A person unable to work because of illness usually receives sick pay that can add up to about 85 percent of his or her normal income. Doctor and hospital care and psychiatric and social counseling are also provided. A wide range of social services are provided for the elderly, including a generous pension system.

The government provides programs that allow families with children to have the same standard of living as those without. Such policies help low-income families the most. The state pays parents a tax-free allowance for each child until age 16. Families with more than three children receive extra aid. If children continue their education after 16, state aid continues.

When a baby is born, the child's parents are entitled by law to 12 months' paid leave, which they can divide between themselves as they wish. But one of the months must be used by the father—and the family loses it if he doesn't use it.

Sweden has made a strong effort to end discrimination based on sex. Over 75 percent of all women between the ages of 16 and 64 are employed. Municipal governments provide day care for infants and after-school care for children.

Education is free for all citizens and permanent residents. Municipal governments provide preschools, which are available to all children under age 7. School attendance is compulsory from age 7 through age 16. Ninety-five percent of all students attend upper secondary schools for two- to four-year courses that prepare them for jobs or for higher education.

Sweden has approximately forty colleges and universities, which are operated primarily by the state. About half the students are women. A large percentage are people over 25 who take advantage of special admission rules for adults with work experience. Tuition is free, and state grants and loans are available to cover costs such as room and board.

The state provides free education, almost entirely free health care, and a wide range of social services to every citizen or permanent resident of Sweden. The price is high, but the standard of living and future prospects for the average Swede are among the best in the world.

A Lapplander family in native costume begins Christmas dinner.

The People and Culture

Swedish culture and society are strongly influenced by Sweden's geography, seasons, and climate. Winter is an inescapable fact. For most of the country, it begins in November and lasts until April. Except for the southern region, the land remains covered with snow during these six months. Lakes, rivers, the Gulf of Bothnia, and in some winters, even the Baltic, freeze solid. And it is dark. Before it became known to tourists as the Land of the Midnight Sun, Scandinavia was known to the rest of Europe as the "midnight land."

The Swedes' first defense against the gloom of winter is light. The streets and houses are brilliantly lit from early afternoon, when the sun goes down, to morning. Even the most remote farmyard is illuminated.

One of Sweden's most joyous festivals celebrates St. Lucia, "the daughter of the longest night." The celebration, held each year on December 13, is a moving sight. White-robed girls are crowned with headdresses of small candles. Attended by other girls carrying candles and boys carrying wands tipped with stars, the white-robed girls take coffee and cakes to the older people of the community.

Swedes decorate their homes for Christmas (or Yuletide, from the old Swedish word *jul* for the Christmas season) with brightly lit trees. Another, briefer light comes during Christmas, when aquavit, a clear, caraway-flavored liquor is poured over cups of mulled wine and set aflame. The warming drink that results is called glogg.

Even though Sweden is the home of the reindeer, Santa Claus is not a native. The traditional Christmas visitor was the *nisse*, or gnome, a short, good-humored sprite with a long beard and a tasseled red cap. Swedes believed him to be thousands of years old. On Christmas Eve, they left him a dish of porridge on the doorstep. The cult of Santa Claus has taken hold in Sweden, however, and Swedish children write to him in his home in Lappland. There is even a Santaland amusement park in the Dalarna village of Gesunda.

Centuries of coping with cold have made the Swedes expert on how to dress during winter. Traditional wool and furs or high-tech miracle fabrics are a mainstay of winter clothing. Every Swedish home has a large closet near the entrance that can hold visitors' coats, scarves, hats, and boots. Swedish theaters and concert halls employ small armies of attendants to handle their patrons' winter clothes.

The Swedes consider the snow a liberator rather than an enemy. Cross-country skiing is almost universal. Every town and city is crisscrossed with maintained ski trails during winter. According to Stockholmites, "There is nothing to stop you from skiing from here to Lappland." Races and rallies fill the winter calendar. One of the most popular is the 56-mile (90-kilometer) Vasa race, held each March in Dalarna. Thousands of participants retrace the route of young men of Dalarna, who raced after Gustav Vasa, skiing to exile in Norway to ask him to lead them in revolution against Denmark.

Swedes love to race and play on the ice, too. A traditional ice game called *bandy* is still popular, and more and more young Swedes are learning the North American game of ice hockey. Another diversion is ice-boat racing. Ice-boats are nothing more than frames on runners that are propelled by sails. When the ice is smooth and long

Thousands of skiers swarm up a hill to participate in the annual "Vasa Race."

enough, and when the wind is blowing right, an ice boat can reach speeds greater than 60 miles an hour (37 kilometers an hour).

A calmer way to enjoy the ice is ice-fishing. Ice fishermen drag brightly painted wooden huts to holes cut in the ice. Inside the huts, warmed by tiny stoves, they drink and snack, chat and smoke, and sometimes even catch a fish.

Hunters head for the mountains and steppes of the north during the winter. Elk are the chief prey. The large herds sometimes damage the forests with overgrazing, and hunting keeps their numbers down. More than 70,000 hunting licenses are issued each year. Hunters pursue wolves in Lappland, because they are considered a threat to the reindeer herds.

Spring is not the most pleasant of seasons. Roads and ski trails become slushy and the fields become muddy. Ice on the rivers breaks up in thunderous roars that warn of coming floods. On the first of May, a national holiday, university students congregate in white caps and radicals parade with red banners, but snow flurries are not surprising, and everyone is still bundled up.

Summer is a three-month celebration. To the great Swedish playwright August Strindberg, it was when "the earth is a bride and the ground is full of gladness." The days gradually grow longer until the climax of Midsummer Eve. Because the sun's rays strike the earth more directly in the north, the light is more intense. Plants grow with abandon, and the smell of flowers is everywhere.

Young Swedes dance around the maypole in celebration of Midsummer Eve.

A young Swedish family enjoys the public beach at Lake Siljan in Rattvik.

In summer, apartment dwellers head for the woods, to open their cottages and collect mushrooms. Others escape to the skerries and launch their boats into the water. Although there are few commercial fishing boats anymore, Sweden has the world's highest per capita ownership of private boats.

In the mountains, bears come out of hibernation, and elk move up to the high meadows. In Lappland, the Lapps herd their reindeer onto the tundra. The Baltic coast and islands teem with migrating birds.

In the cities, parks and gardens are filled with flowers and sunbathers. It seems that everyone eats lunch outdoors. Many businesses shorten the working day, or begin it earlier, so that the offices are empty by no later than 3:00 P.M. In fact, from Midsummer Eve through August, many companies operate with skeleton crews. Swedish workers average five weeks of vacation a year, and most vacation time is taken during the summer.

Summer is the time for folk festivals. One of the prettiest festivals occurs on Lake Siljan in Dalarna. Villagers in their regional costumes cross the lake in 20-oared wooden boats decked with garlands and tree boughs. The villagers dock at a little white church on the shore near Rattvik, called *Den knäböjande bruden vid Siljan*, "the bride kneeling by Siljan's shore."

Famous film director Ingmar Bergman made motion pictures that helped set the tone of modern cinema.

August Strindberg wrote bitter, naturalistic plays that broke new ground in the modern theatre.

Swedes' attitudes toward the state religion, Lutheranism, are complex. The church was once a powerful political force. By law, every citizen automatically becomes a member at birth. Today, the church seems to have little influence on daily Swedish life. Church attendance is fairly low, except at Christmas and Easter. Yet, although anyone who wishes to leave the church can avoid a small tax, more than 95 percent are still members. Most Swedes still turn to the church for marriages and funerals.

Summer is a time for "crayfish parties." Families join together under the moonlight to boil huge batches of the freshwater delicacies, which are eaten to the accompaniment of vodka and laughter. The end of the summer season is usually marked by one last, nostalgic crayfish party.

In the past, the Swedish diet was tuned to the seasons. A year-round constant was fish—salted, pickled, or smoked in winter, fresh from the river or the sea in summer. Potatoes were another staple. For a time the potato was as important to the Swedish diet as it was to the Irish diet. Summertime brought berries, mushrooms, and ferns. In winter, people ate smoked meat and fish and root vegetables.

The famous smorgasbord was once a peasant institution. Whole villages would gather at the end of summer to celebrate the bounty of the land—roast game, boiled potatoes and turnips, fresh, smoked, and pickled fish, meatballs, pancakes, and soups. It was a feast, a ritual, and a way to distribute the surplus of summer before winter set in. Today the term refers to a meal made up of many different dishes, from which the diners choose what they want.

Swedish culture retains its distinctive features, although the spread of global communications has given it an international flavor. Love of culture and art is visible throughout Sweden, in murals, sculptures, paintings, and in the lovely designs of even the most basic tools and utensils in the home.

Swedish literature has not received much international acclaim. For most of Sweden's history, writers relied on French or German writing styles. One absolutely unique voice was Emmanuel Swedenborg's. An 18th-century naturalist, inventor, and visionary, Swedenborg offered a mystical reinterpretation of Christianity that affected many throughout the world.

The most famous Swedish writer is the 19th-century playwright and novelist August Strindberg. Strindberg was a "naturalist" writer;

that is, he portrayed people as being shaped and driven by the forces of nature and society, rather than as idealized or symbolic. His most famous play was *Miss Julie*.

Strindberg's grim, naturalist view of life aroused great controversy, but Selma Lagerlöf's romantic view of life was much more popular. Other "new-romantics" of the early 20th century were Verner von Heidenstam and Erik Axel Karlfeldt. All three won Nobel Prizes, but it is Strindberg's work that attracts attention today. Other, better-known Swedish writers of the 20th century are Par Lagerkvist and Dag Hammarskjöld. Hammarskjöld's posthumous memoir, *Markings*, was a worldwide bestseller.

Swedish writing today is varied. There are many literary journals, and lyric poetry is quite popular. Much current writing is concerned with political questions. Because there is no censorship in Sweden, debates can get quite lively.

Swedes buy more magazines and newspapers per person than any other nationality. The Stockholm daily newspapers are circulated nationwide by air, and every trade group, arts organization, and municipality publishes its own periodical.

Winter or summer, Swedes flock to open-air plays, dance performances, and concerts. State and municipal governments support theatre and dance groups. The Swedish Royal Ballet, under the direction of Birgit Culberg, has become internationally respected.

Music has always been supported by the state. King Gustav III established the Academy of Music in 1771, and the Royal Opera House was built in 1773. Sweden has produced more than its share of great singers, beginning with Jenny Lind, the "Swedish Nightingale," who captivated American and British audiences in the 1850s. In the 1880s, Christine Nilsson sang at the opening of the Metropolitan Opera House in New York. In 1960, soprano Birgit Nilsson made a historic debut in the same house. Several cities have their own symphony orchestras, and chamber music is very popular.

Movies became popular early on. The Swedish film industry has produced a number of world-renowned directors and stars. The most famous was the director Ingmar Bergman, whose brooding, atmospheric films became popular throughout the world. The Swedish actress Greta Garbo became a notable Hollywood star, as did Ingrid Bergman.

More than 95 percent of the homes in Sweden have televisions. The Swedish Broadcasting Corporation (SBC), a government agency,

Jenny Lind, the famed "Swedish nightingale" of the 19th century, made memorable tours of the United States.

controls the television system. No advertising is allowed. The SBC broadcasts an educational channel and an entertainment channel.

Sweden has had several master sculptors. One in particular was Bernt Notke, whose powerful *St. George and the Dragon*, made of wood, stone and bronze, dominates the interior of Stockholm Cathedral. Other noted Swedish artists were the painter Anders Zorn and the sculptor Carl Milles, whose "flying sculptures" can be seen in parks and gardens throughout Sweden and the United States.

Sweden's greatest artistic accomplishments are in the practical arts—the manufacture of beautiful, useful objects. Swedish glass is renowned the world over for its line, grace, and expressiveness. In 1845, the Society for Industrial Design began educating Swedes to appreciate beauty in even the humblest object. The results can be seen today in apartments, public places, and factories. Even the humblest object in Sweden can be a thing of beauty.

A youngster celebrates St. Lucia Day. The ceremony of lights is observed during the darkest days of the year.

Sweden and the Future

In many respects, the Swedes are a forward-looking people. Often the social and political trends that appear in Sweden are clues to what will soon be happening elsewhere in Europe. In the years to come, Sweden will face many challenges, and the ways in which it deals with them are bound to attract much attention throughout Europe.

One current concern of the Swedish people is their environment. Acid rain and nuclear power are two touchstone issues. For years, acid rain has been a serious threat to both the economy and the countryside. Swedes are convinced that the pollutants in acid rain come from the smokestacks of industrial nations beyond their boundaries. They believe that the acid rain is destroying the country's forests and killing its wildlife. Swedish diplomats have worked with neighboring nations to reduce the amount of sulfur dioxide (the main pollutant in acid rain) released into the atmosphere, and Sweden is a party to various international agreements dealing with sulfur dioxide emissions and other pollutants. In the next decade, Sweden is likely to continue leading such efforts.

As for power generation, the Swedish people decided in a 1980 national referendum that nuclear power plants and the waste they produced were too great an environmental threat. Nuclear power was to be phased out by the early 21st century. Now the government has begun to act on that promise, but the details of the transition remain controversial. Some environmental groups fear that the nuclear plants may be replaced by coal- or oil-fired generators, which they see as substituting one form of pollution for another. The business community, on the other hand, worries that alternative power sources may be more costly than nuclear plants and that the transition may discourage foreign investment.

In politics, the major debate is between those who want to continue and expand the welfare state, and those who want to reduce or end state involvement in social welfare. In the election held in October 1991, for instance, the opponents of the welfare state won a major victory. A conservative coalition under Prime Minister Carl Bildt replaced the Social Democrats and began to cut taxes and reduce the role of the government.

Under the conservative coalition, the country slipped into a recession. The conservatives then lost the election held in 1994, but the Social Democrats failed to win a clear-cut victory, falling 13 votes short of a majority in the parliament. The Social Democrats were forced to form a minority government and depend on the support of smaller parties such as the Centre Party. In this tentative position, many observers believed, the Social Democrats were reducing their commitment to the social welfare state.

The policies of the European Union (EU) have added to the debate. In order to participate in a common European currency, EU members were urged to reach stringent goals for their budget deficits and national debt. This would mean trimming many cherished social welfare programs. Opinion in Sweden, as elsewhere in Europe, grew sharply divided about the wisdom of these goals.

One strength of the Swedish political system is the referendum. The constitution allows referendums so that an important issue can be put to a nationwide yes-or-no vote. The 1980 referendum played a major role in settling the future of nuclear energy. Another referendum approved Swedish membership in the European Union in 1994. This supremely democratic technique can help avoid political deadlock in the Swedish government.

The real threat to the welfare state comes from some long-term social trends. The birthrate in Sweden is very low: about 1.7 children per woman. (A birthrate of 2.1 is required to maintain the population at the same size.) Life expectancy, 74 years for men and 80 years for women, is among the highest in the world. As a result of these trends, people older than 65 make up 17 percent of the population. The percentage of retired and very old people is growing higher every year, and the relative number of new workers is not increasing. This means fewer contributions to pension and health-care systems that need more and more money.

One factor that may help is the increase in immigration. Since 1945, immigrants from Greece, Turkey, the former Yugoslavia, and other countries have added nearly a million people to the population. Immigrants have boosted economic productivity and may increase the birthrate. Some fear, however, that people from different lands will change Sweden's cultural and racial identity. Others believe that the immigrants provide new ideas and energies that will invigorate Swedish society.

In resolving its political, social, and environmental questions, Sweden can draw on its unique mix of democracy, government planning, and capitalist initiative. Its decisions will be closely watched, and perhaps imitated, by other nations in Europe and the rest of the world.

This is the entrance court of the Carl Milles Gardens in Stockholm, which displays many of that famous sculptor's works.

GLOSSARY

acid rain	Precipitation polluted with sulfuric acid. The acid is a byproduct of the burning of fossil fuels. Some scientists believe acid rain is responsible for the death of forests and wildlife.
archipelago	A group of islands.
bruk	A small community factory.
esker	A long narrow ridge of gravel and stone left behind a retreating glacier.
fjord	A narrow inlet of the sea between cliffs or steep slopes.
glogg	A holiday drink of mulled wine topped with aquavit and set aflame.
heath	A large, level treeless area of uncultivated land, usually boggy or not well drained, where heather and shrubs grow.
mixed economy	An economic system that is organized on both socialist and capitalist principles.
ombudsman	An official appointed by the Riksdag to protect the rights of individuals and to prosecute officials who violate citizens' rights.
referendum	Submitting an important political or social question to the direct vote of the people.
runes	Characters of the alphabet developed from Latin and Greek letters by the German and Scandinavian tribes between the 3rd and 13th centuries.
Scandinavia	The region centered around the Baltic Sea, con-

sisting of Sweden, Norway, Denmark, and Finland. Because of their shared racial makeup and history, Greenland and Iceland are also considered part of Scandinavia.

skerry A low, rocky islet.

ting The annual meeting of Viking chiefs at which they voted on matters of common concern, such as the election of a king.

tumulus A mound of earth and stone, usually built by prehistoric tribes for the burial of a chieftain or king.

value-added tax (VAT) A tax levied by the government on the value added to a good or service at each stage in the production and distribution of the good or service.

INDEX

A
acid rain 26–27, 111
Age of Liberty 75, 76
agriculture 94
American Civil War 30
Arctic Circle 21, 22, 24
area 21
armed forces 89
armed neutrality, policy of 19, 45, 77, 80, 81, 87, 89
arms manufacturing 17, 30
art 109
Aryans 52
Austria 76

B
Baltic Sea 21, 36, 37
Bergman, Ingmar 108
Bergman, Ingrid 108
Bernadotte, Jean 77
Bildt, Carl 112
birthrate 113
Boat-Axe People 52–53
Bofors, weapons manufacturer 30
Bothnia, Gulf of 21, 24, 25
Bronze Age culture 52

C
Canute, King 44
capital punishment 87
Carlsson, Ingvar 83
Catholicism, Roman 44, 72
Centre party 82, 83, 112
Charles X 44, 73
Charles XII 74, 75
Charles XIII 77
Charles XVI 77
Chernobyl nuclear reactor 19, 92
Chicago 79
Christian II 77
Christianity 44, 66, 67, 69, 72
Christina, Queen 73
civil defense 89
Common Market, *see* European Union
Conservative party 82
Constantinople 56
constitution 75, 82
Copenhagen 44, 74

D
Danegeld 56
Denmark 21, 43, 45, 47, 55, 70–74, 76, 77, 81, 88, 100
disarmament talks 89
discrimination, by sex 97
dolmens 51

E
economic policy 90–94, 112
economy, mixed, 90

education 78, 97
elections 87
energy consumption 91
engineering 94
England 55
    (*see also* Great Britain)
environmental issues 26–27, 111
Eric, patron saint 69
Ericsson, John 29, 30
Erik Eriksson 70
European Union 19, 83, 88, 112, 113
exports 94

F
farming 28
feudalism 70
Finland 19, 21, 25, 69, 71, 75, 76, 77, 88
folk dancing 28
folk festival 104
foreign aid 90
France 55, 76
free-trade pact 88
frontiers 21

G
Garbo, Greta 108
Germany 49, 81
glaciers, impact of 25
gods 52
Gold Coast (Africa), Swedish settlement in 73
Gold Coast (Sweden) 25
Gothenburg 35–37, 45, 72
Goths 22, 28, 55
Great Britain 76
    (*see also* England)
Greece 40, 113

Greenland 56
Gulf Stream 36, 37
Gustav III 76, 107
Gustav IV 76
Gustavus Adolphus 36, 42, 72

H
Hammarskjold, Dag 89, 107
Hanseatic League 70, 71
health care 87, 96, 97
Helsingborg 45
Holy Roman Empire 70, 72
housing 87, 94
human sacrifice 52, 65
hydropower 91

I
Ice Age 25, 51
ice-fishing 101
Iceland 21, 56, 88
immigrants 40, 113
India 52, 90
Industrial Revolution 18, 79
Ireland 56
ironworking 53, 54

J
judicial system 87

K
Kalmar Union 71, 72
kingship, elective form of 70
Kiruna 23, 24

L
labor unrest 79
Lagerlof, Selma 29, 40, 107
Lappland 23, 101, 103
Lapps 23
Liberal party 80

life expectancy 113
Lithuania 56
logging 23
Lund 44, 45
Lutheranism 72, 105

M

Malmö 44
Margaret, Queen 71
Midsummer Eve festival 53, 102, 103
migrations to United States 46, 79
mining 29, 52
Monitor 30
movie industry 108
music 107

N

Napoleon 76, 77
Narvik 24
nationalization 91
NATO, *see* North Atlantic Treaty Organization
New Sweden 73
Nobel, Alfred 17
Nobel Prize 17, 29
nomads 23
Normandy 56
Norrköping 31, 32, 37, 39
North America 56
    (*see also* United States)
North Atlantic Treaty Organization 19, 87
North Sea 21, 24, 25, 36
Norway 19, 21, 55, 71, 72, 73, 75, 77, 81, 100
nuclear disarmament 89
nuclear energy 82, 92, 112
nuclear meltdown at Chernobyl 19, 92

O

Odin 65
oil imports 92
ombudsman 87
opera stars 107
Oscar I 7

P

Palm, August 79
Palme, Olof 19, 82, 83
parliament 85
    (*see also* Riksdag)
paternal leave 96
Peace of Kiel 77
Persia 56
Peter the Great 74, 75
Phoenician traders 54
Poland 73, 74, 77
political parties 80, 81, 82, 86, 112
prisoners, rehabilitation of 87

R

rain 45, 47
referendum 113
Reformation 72
regions 22
reindeer 23
Riksdag (parliament) 39, 71, 75–78, 85
rock carvings 52
Romans 54
runes (writing) 54
Russia 56, 57, 72, 76, 77
    (*see also* Soviet Union)

S

St. Lucia, festival of 99
Scotland 56
shipbuilding 90, 91
Skötkonung, Olof 67

Slavs 56
smorgasbord 106
Social Democrats 80, 81, 82, 83, 86, 112
social welfare 82, 112, 113
Soviet Union 19, 89, 92
    (*see also* Russia)
standard of living 82, 87, 97
Stockholm 30–33, 37, 39–43, 72, 76
Stockholm Bloodbath 72
Strindberg, August 102, 106, 107
Sveas, tribe 25
Swedenborg, Emmanuel 106

T

Tanzania 90
Tartars 56, 69
taxes 82, 83, 86, 90, 95, 96, 97
Thirty Years War 72
trade 49, 53, 71, 80, 88
trade unions 95
transportation 94, 95
trapping 23
tumuli 51
tundra 23
Turkey 113

U

unemployment 78, 90
United Nations 82, 87, 89
    peacekeeping forces 19, 89
United States 79, 89
Uppsala 30, 31

V

vacations 104
Vasa 42
Vasa, Gustav 28, 31, 47, 72, 100
Vietnam 90
Vikings 17, 30, 44, 48, 53, 55, 56, 65, 69
Volvo 36
voter turnout 85
voting age 85

W

wealth, per capita 17
wildlife 26
World War I 80
World War II 81, 87, 92
writers 107

Y

Yugoslavia 40, 113

PICTURE CREDITS

Acme Photos: p. 38; Antikvarisk-Topografiska-Arkivet: p. 54; The Bettmann Archives: pp. 2, 14, 18, 20, 24, 26, 27, 32–33, 34–35, 41, 48, 66, 74, 103, 105, 114; Swedish Information Service: pp. 73, 102, 108, 110; Swedish Tourist Board: pp. 57, 58 (above & below), 59 (above & below), 60 (above), 61 (above), 60–61 (below), 62 (below), 63 (below), 62–63 (above), 64 (above & below); United Press International: pp. 16, 50, 86, 88, 101; UPI/Bettmann Newsphotos: pp. 43, 55, 78, 81, 84, 91, 92–93, 95, 98, 104.